OCT 09 2019

WITHDRAWN
FROM OUR
COLLECTION

153.4
BEC

Beck, Henning

Scatterbrain

HUDSON PUBLIC LIBRARY
3 WASHINGTON STREET
HUDSON, MA 01749
ADULT: 978-568-9644
CHILDREN: 978-568-9645
www.hudsonpubliclibrary.com

SCATTERBRAIN

HENNING BECK

Translated by BECKY L. CROOK

SCATTER

BRAIN

How the Mind's Mistakes Make Humans Creative, Innovative, and Successful

GREYSTONE BOOKS
Vancouver/Berkeley

Copyright © 2019 by Henning Beck
Translation copyright © 2019 by Becky L. Crook
Originally published in Germany by Hanser as *Irren ist nützlich: Warum die Schwächen des Gehirns unsere Stärken sind* in 2017

19 20 21 22 23   5 4 3 2 1

All rights reserved. No part of this book may be reproduced, stored in a retrieval system or transmitted, in any form or by any means, without the prior written consent of the publisher or a license from The Canadian Copyright Licensing Agency (Access Copyright). For a copyright license, visit accesscopyright.ca or call toll free to 1-800-893-5777.

Greystone Books Ltd.
greystonebooks.com

Cataloguing data available from Library and Archives Canada
ISBN 978-1-77164-401-3 (cloth)
ISBN 978-1-77164-402-0 (epub)

Editing by Heather Wood
Proofreading by Jennifer Stewart
Text design by Belle Wuthrich
Jacket design by Belle Wuthrich and Fiona Siu
Jacket illustration by istockphoto.com
Printed and bound in Canada on ancient-forest-friendly paper by Friesens

Greystone Books gratefully acknowledges the Musqueam, Squamish, and Tsleil-Waututh peoples on whose land our office is located.

Greystone Books thanks the Canada Council for the Arts, the British Columbia Arts Council, the Province of British Columbia through the Book Publishing Tax Credit, and the Government of Canada for supporting our publishing activities.

The translation of this work was supported by a grant from the Goethe-Institut, which is funded by the German Ministry of Foreign Affairs.

CONTENTS

INTRODUCTION

THIS IS NOT a book that describes how great the brain works. At least, not at first glance. It's also not a book about how perfectly the brain works. Because it doesn't.

And if, after reading this book, you are hoping to improve your brain's ability to think or concentrate, I'm afraid I have to nip that notion in the bud right from the start. That's not going to happen either because the brain is anything but precise or good at calculating. It's a dreamy scatterbrain, often distracted and unfocused, never one hundred percent reliable; it miscalculates, is frequently inaccurate, and forgets more than it retains. In short, the brain is an approximately 3.3 pound mistake. You carry this head full of sloppy blunders around with you wherever you go—and I would like to congratulate you for that.

Now that I've been permitted to scare most of my readers off, I would like to let you know that there is, in fact, a reason to continue reading this book. I mean to show you that it is precisely these seemingly inefficient imperfections and bloopers that help your brain to be so exceptional and successful.

We are all familiar with this from our own lives. The brain makes mistakes—sometimes big ones, sometimes small ones, and not a day goes by that our brain doesn't concoct some foolishness, misjudgment, or just plain messes up. You underestimate the time, forget what it was you were just reading, or allow yourself to get distracted by your mobile phone. And this is a great thing. Because these supposed weaknesses and imperfections are what make your brain so adaptable, dynamic, and creative.

Do you think I am exaggerating? All right then, let's test your mental abilities:

Imagine you're in a race and you overtake the guy in fourth place. What is your position now?

Third place?

Well, no—but don't worry. Your brain easily creates mental boxes (see chapter 11) and mixes up numbers (see chapter 8). Never mind. Literally. Even simple addition can get complicated. How often is the letter M repeated in the next line?

MMMMMMMMMMMMMMMMMMMMMMMMMMMMMM

Enough hemming and hawing. It's not quite as easy as it seems, is it? What this shows us is that the brain doesn't seem geared toward processing pieces of information that are indistinguishable from each other. On the contrary, it often gets bogged down with such information.

"Making mistakes is what makes us tough, so making just one is never enough," my chemistry teacher once said. And then he proceeded to ignite silver acetylide and blew a crater in the schoolyard. Take note: trial and error is not always the method of choice. Sometimes it is, however, as exemplified by my neighbor. My neighbor is a truly extraordinary character. At two years old, he is already a pretty clever guy with the ability to master things that would bring any supercomputer to its knees. He is able, for instance, to identify his mother's face among a crowd of people and be aware of his own reflection in a mirror. After playing only a single time with a toy car, he knows what a car is. He can point out smoke detectors on the ceiling and thinks potatoes are yummy—tasks which no modern computer could undertake. At the same time, he is constantly making little mistakes. He could barely walk without stumbling a few months ago, his movements are clumsy, his speech fragmented, and he sleeps for more than half of the day—during which time he is completely inoperable. It would be enough to make an engineer put his head in his hands: "What a defective design. Two years old and it still doesn't run smoothly." A bit like a Windows operating system.

Nonetheless, my neighbor is making tremendous progress day by day, at a pace unmatched by any calculator. Every mistake, every imprecision is an incentive for him to try it differently next time and maybe even to get a little bit better. His brain is anything but perfect—and it never will be. Over time, it will of course improve at adapting to its environment, but it will never be immaculate and absolute because it will always retain its ability to err. Only someone who builds mistakes into their actions will be able, at some point, to develop something innovative and new. Whoever attempts to always think as "correctly" as possible, by contrast, puts themselves on the level of a computer: efficient, precise, and speedy—but also uncreative, boring, and predictable.

As adults, we develop an even more obvious form of intellectual drivel. We forget names and faces. We allow ourselves to become easily distracted by WhatsApp messages or lose our plan for the day in the flood of distracting morning emails. We have names on the tip of our tongues that don't ever fully come to us. We misjudge the time as poorly as we do probabilities or numbers. We struggle to choose from among several options. We go blank right before we are supposed to speak in front of an audience. We find it hard to quiet our minds after an exhausting day at work and are the worst at retaining information and learning under pressure.

On the other hand, there is no organ or system, let alone a computer, that is able to solve complicated problems as playfully as we can: 35 x 27 = ?—that's tough without a calculator. But are you able to recognize the latest Taylor Swift song? No big deal, right? Though the above math problem is relatively simple, we can barely solve it in our heads, and yet we can immediately recognize a song, the face of a loved one, or their voice. And we can do this even though it technically takes much more effort to recognize a certain singer on the stage.

It seems as though our brain is particularly bad at carrying out those tasks that are particularly in demand in our current technical and digital world. We want optimization and precision—perfection. But our brain? It does exactly the opposite, eluding this goal. A lot of people imagine how wonderful it might be if the brain functioned like a mistake-free computer. How concentrated, quick, and efficient we would be at solving problems! And it's true: computers don't make mistakes, and if they do, they crash. But brains don't crash (unless there is outside help, but that's another story). This is because the brain works in a completely different way. It is our errors and our inaccuracies in thinking that, in fact, make us superior to computers. All of the horror stories predicting that computers will soon seize the world, leaving us behind in their intellectual shadow, are clearly rejected by biology. In fact, biology seems to contradict the trend of digitalization, that buzzword of our modern times that seems to want school classes networked like businesses, data exchanged and efficiently analyzed. "Classrooms of the Future," "Analyzing Big Data," "Deep Learning"—there is no area of our lives that we don't wish to optimize with the computational power of the machine. The grand ideas of the future, however, will not come from digital, but rather from analog thinking. From brains, not smartphones. Computers might learn things— but we understand them. Computers follow the rules—we can change them.

Computers might be able to beat us at chess or go, but this is neither surprising, creative, nor a cause for concern. I would, however, start to get worried if a computer began making mistakes and then signaled: "Chess? Er, no, I don't really want to actually, that's boring. What I'm really in the mood for is a round of *World of Warcraft!*" Until that happens, the human brain will remain the measure of all things. Precisely because it supposedly functions so poorly.

In this book, I would like to show you what goes on behind the scenes of the probably most erroneous thinking structure in the world (the brain). To describe the way in which the brain uses errors in order to orient itself in the best possible way to its social situation, to think of new ideas, and to generate knowledge. Yes, it makes mistakes in the process, but the paradox is this: it is through our faults and lack of concentration that our most powerful thoughts are generated. Most of our supposed intellectual disadvantages have enormous advantages. The fact that we can't remember names right away is essential in allowing us to develop dynamic memories. Our propensity to be easily distracted helps us to think creatively. And our tendency to arrive late to an appointment because we misjudged the time is a fabulous thing, because if our inner clocks were exact, we would not be able to jump from memory to memory as quickly as we do but would be trapped in a state of static recollection.

Now, this book is not meant to heap exclusive praise onto all of our intellectual weaknesses. Not every error has a silver lining. But being able to recognize why a brain sometimes doesn't function at the push of a button is the most important step toward understanding these weaknesses. It can help us to become more focused in decisive moments, to allow for the free flow of creative ideas, or to better retain memories. The brain is most likely the best example of turning weaknesses into strengths.

PS Ah, yes, like any other product of the brain, this book is also subject to biological weaknesses and is thus not error-free. You've probably already noted a typo or two, misspellings, or numerical errors that may have slipped in. But after this reading, you will understand why this isn't so bad but rather a good thing. As long as it's in moderation. And speaking of moderation, there were twenty-seven M's in a row a few pages ago. If you managed to count the correct number the first time around, you must really have an error-free brain. Which, at times, also isn't such a bad thing.

1

FORGETTING

Why You Won't Remember the Contents of This Book—
Thereby Retaining the Most Important Information

DON'T BE SCARED. I am going to give you a pop quiz right at the beginning of this book. I want to be certain that you, dear reader, are paying attention. Here goes: What were the first three words on the previous page? Okay, that's not a very easy question, no worries. But perhaps you can answer this: What were the first three words of the introduction? And if that's still too hard: What is the title of this book? I'll bet you can answer that. If you responded by saying, "Scatterbrain," well, at least you've shown the strength of ingrained language patterns.

In any case, isn't it astounding? You sharpened all of your senses and focused in order to start reading (at least, I hope so). And yet you cannot remember something you read only two or three pages ago, or else are only able to do so after an intensive round of thinking. Sometimes your thoughts wander or sometimes you are concentrating so much on what you are reading that you forget what you had read only moments before. This will happen throughout the book, regardless of how hard I try to make the text as captivating as possible. As an author, I'm naturally happy if the reader can retain every tidbit that I've laboriously tapped out on the keyboard. But as a neuroscientist, I am also aware that humans rarely remember what they've read. Hardly anyone is going to be able to recall, word for word, the contents of this book by the end. (Although, if that does happen to you, please contact me. Help and the Guinness Book Committee will be on their way.) You should, however, be able to recall the primary message of each chapter. Hopefully. If not, please buy the book again in order to read it from the very beginning, newly cracked open and smelling of fresh ink. That would make me happy, too.

Apparently, the brain is in a permanent state of forgetting. Anyone who has driven a longer stretch in a car knows what I am talking about. You're driving nonchalantly along the road and then you pause after an hour and ask yourself, "Where am I, actually?"—as though you had somehow switched on a mental autopilot that blocked your memory. Who needs a self-driving Google car when our brains have already mastered the art of autonomous driving? The fact that we don't recall much about a journey through the landscape may have one of two causes: Firstly, the surrounding terrain may have been exceptionally boring (anyone who has driven through Kansas on the I-70 knows what I mean). Or secondly, the brain decided it should simply delete most of the information from the previous sixty minutes. The latter case is the default setting of our thought organ.

This isn't such a bad thing when we are driving a car. But there are other situations in which our brains don't notice a lot. What was last night's newscast headline? What were you thinking about last night before you went to sleep? Did you really lock the door? Question after question that the brain does not want to answer. What an incredibly slipshod organ! Always forgetting, deleting, and blundering. But why is that? Why doesn't the brain remember more than it does and why does it seem to wipe out so much information?

Whether it is dealing with banal everyday topics or really important information, the brain discards everything with the same mechanism. In this age of media overkill, we get used to short-term thinking and are constantly bombarded by new information and messages. Articles that we skim but don't retain. News items that we come across in our smartphone apps but soon forget about. Emails that drown in a flood of messages. Never before in history has it been possible to attain so much new knowledge, and never before has it been so complicated to keep hold of what is truly important. What is going on in our brain when we forget something that we just experienced? And what can we do to not to let the most important bits slip from our memory once again?

A dressing room for memories

FIRSTLY, LET ME reassure you: you shouldn't worry too much if you were unable to remember what was written two pages ago. It is not your brain's job to save as much information as possible. Of much more importance is that the brain forgets the right things at the right time, deleting them from consciousness.

Memories are not static; they aren't data bits that can be accessed once the brain has uploaded them. Rather, memories are dynamic and constantly changing. Only in this way is it possible for the brain to generate new knowledge.

In order for this system to continue, your brain has become an expert at throwing things away to keep them from distracting you. The discarded information might be sensory perceptions, as well as memories, new information, or impressions. In order to maintain as flexible and adaptable a recollection as possible, the brain must eliminate as much information as possible. Only that which is very important is allowed to enter our consciousness, ensuring that we may recall it later on.

Although the brain is a very powerful and dynamic organ, possessing in principle the ability to retain much more information than it actually does, it is also quite lazy. This is why it divides up its energies. For this reason, incoming information is not immediately saved in the brain for the long term but is, instead, placed on a trial period.

We know this from our everyday lives, in which we require things to prove their worth before they are allowed to become permanent fixtures in our routines. Imagine, for example, you are in the market for a new pair of jeans. You would never simply grab the first pair you see in the store display window. You would first test them out. So, what do you do? You take the jeans into the dressing room and try them on, paying attention to two factors: Do they fit well? And do they match your style?

The brain essentially does the same thing. Well, okay, not exactly the same thing, since our heads are more complicated than garments. But the principle is similar: before we decide to commit something to long-term memory (that is, available after several hours or days), it has to pass through a trial period.

Our intellectual dressing room is the hippocampus, a banana-shaped structure located in the center of the brain between our two cerebral hemispheres. Because the first neuroanatomist to describe this structure believed it resembled a sea horse, he named it the hippocampus (the Latin term for sea horse). I wonder sometimes what drugs my colleague must have been taking because I myself have never seen anything in the shape that resembles a sea horse; it doesn't even look like a snake or an eel or any other kind of marine animal to me. To me, the supposed sea horse looks much more like a banana-shaped C smack dab in the middle of the brain.

Each half of the brain possesses a hippocampus that helps us save short-term memories. Everything that should be saved in long-term memory is first "tried on" in the hippocampus. Quite like checking whether the jeans fit you well, the brain also decides whether a possible memory goes well with your previous wealth of experience. The corresponding information is therefore stored in the hippocampus, a process lasting for a few seconds (though if you are hit on the head during this critical phase, your short-term memory will also be gone) or a few hours. The hippocampus will retrieve and analyze the information later—at the very latest while you sleep—in order to decide whether the information should be saved long-term. The decisive criterion is how novel the information is. It is only when something truly new happens to us, which promises to benefit us in some way and that clearly stands out as divergent from our previous experiences, that we will "purchase," or rather, save the information. This transaction costs something too—namely, energy which our neurons must produce in order to adjust their synapses to create a long-term memory. Energy expenditure is the reason why the brain is cautious about remembering. Only the most valuable information is retained; almost everything else is forgotten—even if it's something we see all the time.

A bite in the apple—right or left?

WHAT SHAPE IS the Apple logo? You probably know right off the bat: a bitten apple, black against a white background. But, is the bite mark on the right or the left side of the apple? Does the apple have any other bulges or concavities?

The Apple logo seems very familiar to us because we see it all the time, but in a study conducted through the University of California in Los Angeles, only one of eighty-five participants was able to correctly draw the logo on the first attempt (these test subjects even lived in the country of Apple's origins), and less than half of them were able to select the correct logo from a selection of slightly varying logos.[1] It's no wonder then that it's so easy for copycats to rip products off. On that topic, a little tip for all you vacationers looking for a good deal at the beach: "Guchi" is not written with a "ch."

The more often we are confronted with a piece of information, the duller our memory of it becomes. It is not only the Apple logo that we filter out over time. Study participants have also found it nearly impossible to recall the critical locations of fire extinguishers,[2] the order of characters on a computer keyboard,[3] or the exact details of transportation signs.[4] Do you perhaps know how many people are depicted on a standard pedestrian crosswalk sign? Our brain does not function as a memory machine designed to save details. Instead, it is equipped to forget every last little detail or, to put it another way, to sacrifice to the greater good for the bigger picture.

Active forgetting

SO FAR, SO good. Our intellect filters out reoccurring sensory impressions and sends them into our subconscious. The tiny (and mostly insignificant) details of our memory are sacrificed for the purpose of seeing and recalling the big picture later on. But sometimes one actually does wish to take note of something or other but finds that it has vanished almost immediately from memory—for example, a newspaper article that you just finished reading. You peruse the article only to realize at the end that you have hardly retained any of the information. Or you have just finished watching a daily news program on TV and try to jog your memory about the whole lineup of news stories (which is no easy task, by the way). In these cases, the brain seems to be applying its filter to information that is clearly useful.

Don't worry. This is not detrimental, rather it points to the brain's original strength. Because how relevant is it for us, ultimately, to be able to recall all of the little details of our lives? It is much more important for us to be able to recognize larger patterns from the news and the daily bombardment of information to which we often subject ourselves. In order for us to be able to retain valuable pieces of information, our brain has to forget in a manner that is both targeted and controlled.

Can you recall, for example, your very first day of school? You most likely have one or two noteworthy images in your head, such as putting your crayons and pencils into your pencil case or the first time you went into the classroom. But that's probably the extent of the specifics. Those additional details that are apparently unimportant are actively deleted from your brain the more you go about remembering the situation. The reason for this is that the brain does not consider it valuable to remember all of the details as long as it is able to convey the main message (i.e., your first day of school was great). In fact, studies have shown that the brain actively suppresses regions responsible for insignificant or minor memory content that tend to disturb the main memory.[5] Over time, the minor details vanish more and more, though this in turn serves to sharpen the most important messages of the past.

Rather than allowing intricate details to dim our memories, the brain also deletes these patterns of activity, sacrificing them to the greater good for a somewhat abbreviated but also sharper memory of the main event. Thus, if you wish to retain a detail-rich memory of the past, your best bet is to recall your memory as infrequently as possible. Of course, you won't get much out of the memory because you won't actively be remembering it. But at least you could comfort yourself in knowing that your detail-rich memories have not yet been actively deleted and are still floating around somewhere inside your head.

An intellectual bookmark

AS IMPORTANT AS it is for the brain to forget actively in order to accentuate valuable information, it is equally important that significant information is set aside for later use. Even if you can no longer remember what was in yesterday's news stories, the informational content has not yet been forgotten. You simply cannot recall it—that is the difference.

What does that mean exactly? When we see or hear something new, we don't know right away whether it is going to be important later on. Therefore, the brain has to tag the kind of information that may be used later on so that it might more readily recall it in the future. Think of it as an intellectual bookmark of sorts. We do this in our houses or apartments too. Various objects are scattered around, some of them maybe not so valuable or useful at first glance. We could throw them away, but then we consider they might end up being useful at some point... so we decide to hang on to them. We collect these objects in boxes and baskets and store them away in the attic. And we don't even really remember what we have up there (we've most likely forgotten). But if a golden opportunity opens up in the future, we can dig out the objects and put them to use.

This is what the memory is like. Of course, our brain doesn't store everything in intellectual boxes or baskets, but it does use a similar technique to bookmark potentially valuable information for the future. For the short run, however, the information can be deleted from our conscious memory. In order to demonstrate this, a study tested the bookmarking behaviors of participants.[6] First, they were shown pictures of tools and animals. A few minutes later, the participants were again shown images of either tools or animals, but this time they also received a small electrical shock whenever they looked at the animals. It's no surprise that it was much easier for these participants to later recall the images of the animals, which had been accompanied by an electrical shock! On the following day, the participants were able to list off several of the animal images which they had seen even *before* the electric shock had been administered. It was as if the subsequent electrical shocks had helped the test subjects to dig out their earlier memories even more efficiently. How practical! Finally, a scientifically proven method to kick-start the memory: electroshocks at the right moment can work wonders.

But before you run off to the nearest self-defense store to purchase a memory-jogging device—wait! A radical method such as this is only the second-best solution. A much more important piece of advice is this: even when you seem to have forgotten things from your past, your brain is able to dig them out—when they become important. Very little information is actually deleted permanently. Most of it exists in a waiting state. The brain's supposed weakness (namely, that it quickly blocks out and seems to forget so many things) turns out to be its strength, enabling it to kill two birds with one stone. Firstly, the brain avoids getting bogged down with too much information. And secondly, this enables it at a later time to more flexibly select which information should be remembered. If the brain had to decide immediately which new information, and in which context, should be stored long-term, it would be too sluggish. We are only capable of building up new knowledge when our memories are unstable.

The intellectual tax return

IT SOUNDS LIKE a paradox to claim that the brain is able to produce new knowledge precisely because the brain is so bad at retaining information accurately. The way our memory is organized seems to go against our everyday experience. If we want to organize something in real life, we do it in a particular location. We save our tax documents and receipts in a certain folder, which we place into a cabinet where we can easily find them later on. We put a receipt for a business meal into a folder labeled "Additional Expenses" (if the deal was a good one), and that is how we create order, avoid chaos, and work productively.

The brain is theoretically equipped to do the same thing, to store information in a way that is spick-and-span, orderly and efficient. But it doesn't. If it did, it might be able to master its forgetfulness, but the brain would thereby lose one of its greatest strengths in the process—namely, its ability to dynamically combine information. If you sort your information too early on, it's much harder to put things into a different kind of order later down the road. This pinpoints the difference between a brain and a computer. Whereas a computer mindlessly saves information, the brain creatively combines it to make something new.

Thus, if you were to ask your brain to file an intellectual tax return, it would never sort the business meal receipts into detailed folders but would first put them all into a single stack and mark them each in different ways. You could use the business meal receipts to find out a variety of things. You could review whether a certain restaurant was too expensive, what exactly you ate, or what your client enjoyed eating. This manner of flexible organization only works, however, if you do not determine too early on how the information should later be used, allowing you in retrospect to decide what to do with any given piece of information.

The benefit of shaky memories

THE ABOVE MIGHT sound strange but scientific studies have confirmed it.[7] Test participants were first asked to memorize a list containing words from four different categories (furniture, modes of transportation, vegetables, and animals). Shortly thereafter, they had to learn a typed keyboard combination by heart. Unbeknownst to them, the order of the combination followed the pattern of the word categories (a piece of furniture corresponded to the typed number 1, a mode of transportation with number 2, a vegetable with number 3, and an animal with number 4). The list of words and the keyboard combination both followed the same basic structure, so it was no surprise that the test subjects were able to learn the keyboard combination with remarkable speed, since it matched the word list form they had memorized earlier. What was interesting, however, was that in a follow-up test twelve hours later, the subjects' ability to type in the correct key combination improved the more they had forgotten the list of words—as though the word scheme had been directly "copied and pasted" onto the scheme of keyboard keys.

You are doubtless already familiar with this scientific hypothesis: the more insecurely we save and store a piece of data, the easier it is to combine it with other things. Every piece of information that has not yet been fixed into our memories finds itself in a strange state. It may interact with other impressions and input and influence the learning process. Of course, because the memory must be unstable and shaky, information may also be more easily lost in this state.

In order to gather new knowledge, we are therefore compelled to forget concrete details. But forgetting details isn't such a bad thing since, first of all, the enormous mass of corresponding details would eventually overwhelm even the best brain. And second of all, details really aren't that important. We pick up on patterns, abstract correlations, and the stories behind them—not the little things that often only serve to trip up the brain. In other words: forgetting is a means to an end.

Mental digestion

RECENT RESEARCH HAS shown there is one thing that the brain especially requires to fulfill all of its functions: taking breaks. This is particularly a problem in our modern era, where we are inundated with news headlines, articles, phone calls, and emails. As soon as our brains receive a new piece of information, another piece of information comes along to compete with it. Under such conditions, it's hard for us to evaluate (and forget) individual memories in order to build up new knowledge.

This is why, at this point, I am going to say: don't overstrain your brain's filter—and forgetting—system. Instead, make sure to give it breaks and rest at regular intervals. Because we don't learn when we think that we are learning. We learn in the pauses *between* the thinking. Just as athletes don't improve during their training, but rather in the rest periods between training sessions when they allow their bodies to adapt and heal.

When I read the newspaper over breakfast in the morning, I don't look up the most recent news stories on my smartphone afterward while I'm on the subway. Instead, I wait. I allow myself to get a little bored. This requires a fair amount of courage, as anyone knows who commutes on the subway and who is not absorbed on their smartphone. You start to feel a bit like a communication relic from the 1990s, locked out of the modern Apple and Android universe. But I know it's worth receiving pitying glances from fifteen-year-olds who just earned their latest Candy Crush levels on their phones.

I know I won't be able to remember all of the details I read in my morning newspaper over breakfast. But just as my gastrointestinal tract is digesting my muesli and breaking it down into individual components from which my body will later generate new cells (hopefully a lot of muscles and as few fatty tissues as possible!), my brain is also, in this moment, breaking down the pieces of information from my morning. I can no longer taste the muesli in my stomach any more than the newspaper articles are still present in my mind. But they are having an effect on my brain. And depending on how my day goes, my brain will dig out some or other piece of news from earlier in the day, combine it with the present moment and allow me to brag about my knowledge (which I really enjoy doing). This is only possible if I have taken sufficient informational breaks to allow for mental digestion.

Forgetting in order to retain

NOW YOU KNOW why we (seem to) forget so many things in life. Either they are so homogenous that our brain's information filters cannot distinguish them, or else they are so important that they first have to float around in an unsorted limbo in our subconsciousness so that they can later resurface to combine more flexibly with other bits of information. Strictly speaking, you don't actually forget about these things. You simply aren't able to recall them for the time being. Don't underestimate the extent to which your brain, without your conscious effort, is able to recognize patterns and relationships in your life. You may not be able to remember a particular conversation with your boss, but your brain retains the really important content for when you might need it later.

All of this is only possible, however, if you don't subject your brain to information overkill and the constant bombardment of new data. If your brain remains in an overloaded state, it wouldn't be able to pay attention to the contents of new information, but only to how it changes (rings, vibrates, buzzes, or pops up on the screen). Your brain would eventually set its filter threshold so high that very few new pieces of information could be consciously experienced. You can avoid this by taking deliberate breaks and giving your brain some downtime to reflect.

And now: a break!

CAN YOU REMEMBER the first three words from two pages ago? You don't have to. It's not important because forgetting details is one of the brain's methods. Detail forgetting allows the brain to work its magic of recognizing patterns. It's the same with this chapter. If you are able to remember that it's not a weakness of the brain to forget something, but rather a clever trick of choosing the most relevant bits out of a jungle of information and later combining them in new ways, you've successfully grasped the most important message. The brain is neither a memory machine nor an organizational fanatic that goes around pedantically making sure nothing is forgotten and everything is neatly in its place. No, it's much more scatterbrained than that, bouncing around from one thought to another. But these leaps of thought are precisely what make us creative and independent.

Even though you are probably going to forget most of the details from the past few pages, please try to hold onto this correlation, which is the most significant one: breaks make it possible for your brain to organize information and to bookmark it for later use. Give yourself the freedom to put down this book for a few minutes. Relax a little and let the information soak in before you continue to read. Because now you know that, even if you can't remember the chapter, your brain is diligently taking note of its most valuable information for later on.

2

LEARNING

*Why We Are Bad at Rote Learning, but Better
at Understanding the World*

KNOWLEDGE IS POWER, everyone says, and so it follows that the most powerful people must also have the most knowledge. But, it turns out they usually have the least. Knowledge doesn't simply rain down from the sky. Our brain has to work to attain it; it has to learn. And this isn't a very easy task. Try it out right now by memorizing the following list:

Ginger
Raisin
Bicycle
Strawberry
Night

Hedgehog

Salad

Grapes

Noodles

Clock

Rest

Dream

Zebra

Lollipop

Labyrinth

Chameleon

Raspberry

Allow yourself to read the list multiple times so that you can really get it. Feel free to use tricks, imagery, mnemonic devices, storytelling. Then continue reading. But don't forget: don't forget! Even though the previous chapter showed us just how hard it is to remember things and also that the brain loves to toss things out of its memory.

Learning isn't everything

THE IDEA OF learning doesn't conjure the rosiest image for us. This is apparent in the words that we use around the idea of learning: cram, bone up on, wade through, bury oneself in, burn the midnight oil, or even put our noses to the grindstone. A lot of people associate learning with an unpleasant period spent at school or attending a training course accompanied by exertion, frustration, battles over grades, and annoying exams. Life is divided into the time in which one is required to learn, finish homework or seminar materials, and spare time, in which we can finally do something fun. Learning is tedious, exhausting, and undesirable. Spare time, free from learning is, by contrast, fun, relaxing, and enjoyable. It almost seems that we need to create a special environment for learning if it is ever to happen at all. Anyone who wants to continue their education has to take a course or a workshop, and when it's finally over, they have "learned enough." The exam is passed; the certificate is in hand—it's curtains on learning.

Unfortunately, learning doesn't simply let go quite so easily. We are continuously required to educate ourselves and there is never an end to it. I recently read in my autograph book a reflection written by a then seven-year-old pal who had already figured out, over twenty years ago, that he was never going to stop learning: "Learning is like paddling against the stream. As soon as you stop, you float backwards." The buzzword nowadays is "lifelong learning." And of course, we do have to learn everywhere and all the time—at school, at university, in our careers. We are thus fortunate to have a brain that learns with us.

Or, does it? At the end of the day, it is not very easy to acquire and save information. In fact, it turns out the brain has three weaknesses when it comes to learning. The first is that it doesn't learn very well under pressure. Anyone who has ever studied for an important exam knows how complicated that can be. Secondly, we are extremely bad at learning data, facts, and information. The brain tires quickly of this kind of stuff. Or are you perhaps able to recall the names of the first five Presidents of the United States, the second binomial formula, or the difference between a predicative and adverbial clause? No? You have probably learned all of these things at some point but then forgot them again. Which leads us to the third of the brain's weaknesses: anyone who is able to learn something is also able to *unlearn* it. Learning is not a one-way street of knowledge in the brain.

Although at first glance learning appears to be a tedious business, linguistically disparaging, and an altogether arduous undertaking, the brain happens to be a grand master in this particular discipline. After all, learning is our evolutionary specialty, our ecological niche—the thing that we are able to perform with exceptional agility and which sets us apart from other species. Birds fly. Fish swim. Humans learn. Albeit differently than we might suppose. There's no doubt we have certain weaknesses when it comes to learning (i.e., the stress of learning causes us to cramp up, we are bad at memorizing facts, etc.) but on closer inspection, it becomes apparent that these deficiencies are merely the price we pay for being the best learners in the world. Or, even more than that: not only do we learn, we also understand the world. This is the great strength of human thought and why it is worth swallowing the few weaknesses that go along with it. Anyone who is able to appreciate this should also be able to understand the best methods for taking in new information (how best to "learn") and why we, as a species, will always remain superior to computers.

The neuron orchestra

BEFORE WE START talking about the weaknesses (and strengths) of how we learn, we'd better take a peek behind the scenes of a learning brain. What is it that happens, in fact, when we learn something new? Or, we can ask an even more basic question: What is a piece of information—is it a thought inside of our head that needs to "get learned"?

When it comes to computers, the answer is relatively clear. If I want to save something on a computer, I first need something to save. We call this data, electronically processed characters. The computer has to put these bits of data somewhere so that it may obtain them later. It organizes a data packet, or a location where it can selectively access the material. Once it has both (data and location), the computer is ready to process this combination as information. This is not unlike what goes on in a library. The books contain (written) characters that are placed on the shelves in a system that helps you to locate them again. If you want to get ahold of a piece of information in a library, you will need both components here as well. You will need to know where the book is located, and you will need to be able to process the characters in the book.

It's different in the brain, however, because there are neither characters (data) nor a fixed location where the data is held. If I were to say: "Think about your grandmother!"—you wouldn't get some kind of "grandmother neuron" suddenly popping up in your brain (as brain researchers used to believe)—instead your neuronal network would assume a very particular state. And it is precisely in this state, in the way in which the nerve cells activate each other, where the information is located. This may sound somewhat abstract, but let us simplify it by comparing it to a very, very large orchestra. Individual members of an orchestra can also individually change their activity level (playing louder or quieter or at higher or deeper tones). If you are watching a silent orchestra with inactive musicians, it's impossible to know what compositions they have in their repertoire.

In the same way, it's impossible to know what a brain is able to think by simply observing it from outside of its neural network. In an orchestra, the music is produced when the musicians play together and in sync. The music is not located somewhere within the orchestra but is rather *in the activity* of the individual musicians. If you only listen to a single viola, you can gain some insight into one musician, but you won't have any idea of what the complete musical piece sounds like. In order to know this, you also need to find out the way in which the other musicians are active at the same time. But even this would not be adequate because in this case you would only know what one particular tone sounds like at any one given moment in time, whereas the music only first emerges when you consider it over the course of time. The information (in this case, the melody of the musical work) is located *between* the various musicians.

Like orchestra musicians, neurons also tune themselves to one another. Just as an orchestra produces a piece of music when the musicians interact, neurons produce the informational content of a thought. A thought isn't stuck somewhere in the network of a brain. Instead, it is located in the manner in which the network interacts or plays together. In order for this to go off without a hitch, the neurons are connected to each other over common points of contact (synapses), which is the only way that the individual nerve cells can figure out what all the others are up to. In an orchestra, every musician listens to what the others are playing to ensure that they can keep in sync and in tune with each other. In the cerebrum, neurons are connected with several thousand other nerve cells, which means that they are able to produce much more complex states of activity than an orchestra. But it is precisely in these states of activity that the content of the brain's information is located. In an orchestra, this is the music; in the brain, it is a thought.

This method for processing information has a couple of crucial benefits. Just as the same orchestra is able to play completely different pieces of music by synchronizing the playing of the individual musicians in a new way, the exact same neural network is able to produce totally unique thoughts merely by a shift in activation. In addition, a piece of information (whether a melody played in an orchestra or an image in one's head) is not necessarily coded in a concrete state of activity, but also in the shift of the state. The mood of a piece of music may be influenced by whether the musicians play softer or louder—in the same way, the information in the neural state may also be influenced by the way in which the neurons *shift* their activity and not only how they currently *are*.

This brings us to the realization that the number of possible patterns of activity is vast. The question of how many thoughts it's possible to think is thus as useful as the question of how many songs it's possible for an orchestra to play.

There is something else to notice here. In a computer, the information is stored in a location. When you switch the machine off, the information is still there (saved in the form of electrical charges), and all you have to do is to turn the computer back on to retrieve it. But if you switch off a brain, the party is over. End of story. Because the information stored in a brain is not located in any particular physical location but is rather an ever-changing state of the network. During a person's lifetime, a thought or a piece of informational content always proceeds from an earlier one—as though every state of thought becomes the start signal for the next thought. A thought is never derived from nothing.

The learning in between

AS USEFUL AS the orchestra metaphor is, I don't want to conceal the fact that there is one enormous difference when it comes to the brain. And the difference is this: unlike an orchestra, the brain does not employ a conductor (and the neurons also don't have predefined sheet music to play). There's no one standing on a podium in front of the neurons to direct them on how they should interact with their neighbors. And yet they still manage, with utmost precision, to synchronize themselves in their activities and to create new patterns.

This has consequences for the manner in which a neural network learns. While an orchestra conductor provides the tempo to sync up the musicians, the neurons have to find another method. And as it turns out, the way that information is produced is somewhat like the *melody* of an orchestra, in the ability of the individual neurons to play all together.

When an orchestra learns a new melody, the musicians must accomplish two things. Firstly, they have to improve their own playing skills (i.e., learn a new combination with their fingers). Secondly, and also more importantly, they have to know exactly when and what to play. They can only be really certain of this, however, by watching the actions of the conductor and waiting to hear how the others around them are going to play. When an orchestra practices a new piece, the musicians are in effect practicing their ability to play together. At the end of the day, the piece of music has also been "saved" in the orchestra's newly acquired skill of playing it together. In order to retrieve it, the concrete dynamic of the musicians first must be activated, leading to the piece of music. Likewise, a piece of information in the brain is encoded in conjunction with the interaction between the neurons, and when the neurons "practice," they also adjust their harmonization with each other, making it easier to trigger their interaction next time. In order for a neural network to learn, the neurons must also adjust their points of contact and thereby redesign the entire architecture of the network.

Because the brain does not have a conductor, the nerve cells must rely on tuning themselves to their neighboring cells. What happens next on a cellular biological level is well known. Simply put, the adjustments among the neural contact points that happen during learning follow a basic principle: contact points that are frequently used grow stronger while those seldom put to use dwindle away. Thus, when an important bit of information pops up in the brain (that is, when the neurons interact in a very characteristic way), the neurons somehow have to "make a note" of it. They do this by adjusting their contact points with one another so that the information (the state of activity) will be easier to retrieve in the future. If in specific cases, some of the synapses are quite strongly activated, measures are taken to restructure the cells to ensure that it will be easier to activate the specific synapse later on. Conversely, synapses that go unused because of a lack of structural support are dismantled over time. This saves energy, allowing a thinking brain to function on twenty watts of power. (As a comparison, an oven requires a hundred times as much energy to produce nothing but a couple of bread rolls. Ovens are apparently not all that clever.)

This is how the system learns. By altering its structure so that its state of interaction can more readily be triggered. In this way, the piece of information is actually saved in the neural network—namely, "between" the nerve cells, within their architecture and connection points. But this is only half of the story. In order for the piece of information to be retrieved, the nerve cells must first be reactivated. The more interconnected the points of contact are, the easier it is to do this, even though information cannot be derived from these contacts alone. If you cut open a brain, you will see how the cells are connected but not how they work. You won't have any idea what has been "saved" in the brain, nor what kind of dynamic interaction it could potentially produce.

Under stress, learning is best—and worst

THIS NEURAL SYSTEM of information processing is extremely efficient. It is much more flexible than a static computer system, requires no supervision (such as a conductor) and, in addition, is able to adapt to a vast range of environmental conditions. However, this learning method also has its weaknesses. Because the process of building neurons is subject to regular biological fluctuations, we don't always learn equally well. When we are under stress, for example, we tend to tighten up more readily. Anyone who has felt the pressure of studying for a test knows how hard it is to prepare with this kind of learning stress. It feels like an arduous task to try jamming the most important bits of information into your head. Or, if you do manage to squeeze them in, you then can't seem to get them back out at the crucial moment (during the test). Why does stress affect our learning process so negatively?

First of all, let me give you the good news: stress is not something that blocks our learning. On the contrary, stress is actually a learning accelerator. Under acute stress conditions (for example, if we are scared or even positively surprised), the brain's neurotransmitter noradrenaline first makes sure to activate precisely those regions of the brain that heighten our attention.[1] About twenty minutes later, this action is further supported by the hormone cortisol, which silences the distracting background flurry of nerve cells.[2] We are then able to become more focused and concentrate. The conclusion? Under acute stress, we are extremely capable of learning. For example, if we cross the street and almost get hit by a car, we take note of this for future street crossings. This is even the case when we are positively stressed. For instance, most of us will never forget our first kiss—even if we only experienced it once.

When our brain is under stress, our neural network is animated, enabling us to learn more rapidly. However, if the content that is being learned does not have anything to do with our stress, it's a very different story. The main goal of a brain under stress is to concentrate only on the stress-related relevant information. Everything else becomes unimportant. And this is what makes learning under stress a two-edged sword. When test participants are placed under stress conditions by having their hands submerged in ice water for three minutes while simultaneously tasked with memorizing a list of words, a few days later, they are readily able to recall all of the words having to do with ice water (such as "water" and "cold"), but they cannot remember any of the other arbitrary words ("square," "party").[3]

If you are almost hit by a car, you are able to draw an immediate correlation between looking both ways before crossing the street and possible death. And you will never forget it. But if you are studying Latin vocabulary, you have to stretch your imagination three times as much to establish a connection between the phrase *"alea iacta est"* and the consequences of bad test scores.

A brief interim conclusion at this point: the brain learns quite well under stress *if* the main point of the learning has to do with the cause of stress itself. After touching a hot stove only once, we are quick to learn that it was not such a good idea. Stress hormones actively regulate the dynamic of the neurons in order to better retain emotional content (the pain from a hot oven is much more important than what brand of oven it is). This is all about emotions, by the way, not facts.[4] Facts, facts, facts are boring. Which leads us to the next learning weakness of the brain.

The memorization weakness

DO YOU REMEMBER the list at the beginning of the chapter? Can you recall even half of it? If yes: I owe you congratulations and respect. How did you go about memorizing this list? If you used mnemonic devices, storytelling, or images to help you to relate the various words, did you realize that you actually increased the amount of information that had to be learned? You made yourself "learn" more than was necessary in order to retain the information. This is a paradox. You might ask an additional question: Why does any of it matter at all? The words on the list are mostly arbitrary and have no relevance or context to you. Why should you be bothered to learn them? Merely because an author demands it of you?

This is precisely the point. Our brain is good at adapting to many different situations, actively adjusting itself, and learning new things, but this doesn't include raw information such as a few random words, bits of data, or facts. Research shows that the upper limit of objects that can be memorized (without using memory tricks such as mnemonic devices or storytelling) is around twenty. Which is not very much. The list at the beginning of this chapter only takes up 146 bytes on a computer hard drive, while a picture of a zebra could easily take up a million times more space. And yet we prefer to imagine, as in a *dream*, a *zebra* with a *lollipop* wandering through a *labyrinth* (words from the list) instead of learning each of these four words separately. But why is the brain so bad at saving a few simple pieces of information, such as a couple of words?

The reason once again has to do with the way the brain works. The brain doesn't learn information by rote and then save it somewhere. Instead, it organizes knowledge. There's a difference. Let me give you a simple example to illustrate. I could list off to you the exact sequence of goals (and who scored them) from the historic semifinal in the 2014 World Cup soccer game between Germany and Brazil that ended in a score of 7–1: *11th minute: 1–0, Müller; 23rd minute: 2–0, Kroos; 24th minute: 3–0, Kroos...* Okay, I'll spare any Brazilian readers the rest of the list and get to the point. Once you have assembled all of the data from this game, what do you know about the game itself? Not much, since you are not witnessing the shocked expressions of the Brazilian team or the joy of Philipp Lahm, the German team captain. The significance of the game cannot be derived merely from the combination of data. Only after you have watched the game can you understand why the Brazilians still grumble about it—in spite of their later "revenge" on Germany in the 2016 Summer Olympics.

Massed learning

UNFORTUNATELY, MANY METHODS for learning (whether in high school, at the university level, in vocational training, or continuing education at the workplace) continue to rely on the basic concept that memorizing facts and data is a good idea. On the contrary, this method leads to a completely false strategy for learning, known scientifically as "massed learning," in which you must pump yourself full of information in a short period of time in the hopes of retaining as much as possible in the future. This obviously doesn't work, since our brain thinks data packets are totally uninteresting.

An orchestra doesn't learn a new piece of music merely by playing a single note for one second and then waiting before processing the next informational packet (the next note) and so on for thousands of notes (this would be akin to "massed learning"). No, it learns best by quickly recognizing the relationship between the notes and the way in which, at a certain time and place, they develop into a whole melody.

The context is what allows us to learn effectively—namely, that we do not have to consciously concentrate on the idea that we are learning. This became evident in a study conducted by the work group of my colleague, Melissa Vo, who researched memory capacity among adults. Specifically, the study's adult test subjects were asked to find objects that were pictured in an apartment setting (i.e., the soap in the bathroom). Although participants were not asked to remember these objects, they were much better able to recall them later than if they had been asked to memorize isolated pictures of the objects.[5] When the same objects were isolated and presented to participants in front of a neutral background, the information was much less interesting to participants and thus not saved. A bar of soap makes much more sense in the context of a bathroom than surrounded by a green background. The object by itself is not interesting. It is only its situation in a particular context that gives the object a meaningful correlation, which we do not forget. Though this may seem illogical because it implies that we are required to note down additional information (namely, the object's surroundings), this is, in fact, an ability that comes easily to us.

The lasagna-learning rule

IN ORDER TO understand this correlation of context and of the meaning of a word, the brain must learn differently than it might be used to—namely, with interruptions. In the last chapter, you already read that the brain sacrifices some pieces of memory to nonmemory (or even actively forgets them) in order to be able to actively combine them. Something similar takes place with learning. Learning is successful when breaks and distance are built into the process, a practice referred to as "spaced learning." This would seem to go against our better intuition, as we assume that we will only be able to grasp correlations and concepts by processing as much information as possible at one time. If we deliberately incorporate breaks into our learning process, we fear we might forget things that could be important. But our brain is not interested in the sheer mass of information so much as it is interested in our ability to connect the information.

To research this, one study asked participants to identify the painting style of various artists. The subjects were divided into two groups. The first group was shown a series of six images, all of them works by one artist, followed by another series of six images, which were by the next artist, and so on for the next four artists. The second group was shown all of the images mixed up in no particular sequence, so that the various artistic styles alternated from image to image. The results were clear. The group that viewed the alternating images was able to identify a new image according to the particular style of the individual artist. Those in the first group, who viewed the images in sequential blocks, were less able to recognize the basic painting concept (artistic style). Despite the results, most of the test subjects indicated that they preferred learning in blocks ("massed learning"), as they believed it to be a more successful strategy.[6]

This result has been reaffirmed over and over again in studies. Taking breaks is what makes learning successful. Not only for learning about various artistic styles, but also vocabulary at school, movement patterns, biological correlations, or lists of words. The reason for this has to do with the way in which our nerve cells interact. An initial information impulse triggers a stimulus for structural change in the cells. These changes must first be processed to prepare the cells for the next informational push. Only after they have taken a short break are they optimally prepared to react to the recurrent stimulus. If it comes too early, it will not be able to fully realize its effect.[7] It is only by alternating information that the brain is able to embed it in a context of related bits of knowledge. It's not too different from making lasagna. You could of course choose to pour the sauce into the pan all at once and then pile the lasagna noodles and the cheese on top. That would be something like "massed cooking," but it wouldn't result in authentic lasagna. Only when you alternate the components do you get the desired, delicious dish— or, when it comes to the brain, a meaningful thought concept. This kind of conceptual thought is the brain's great strength because it enables us to get away from pure rote learning. Only then is it possible for us to organize the world into categories and meaningful correlations and, thereby, to begin to understand it.

Don't learn—understand!

ANYONE WHO CAN learn something can also unlearn it. But once you have understood something, you cannot de-understand it. Learning is not particularly unique. Most animals and even computers can learn. But developing an understanding of the things in the world is the great art of the brain, which it is able to master precisely because it does not consume and draw correlations from data in the same way a robot would. A brain creates knowledge out of data, not correlations. These are two vastly different concepts, though they are often equated with each other in the modern, digitalized world. But whereas the amount of data from :-) and R%@ is the same, the information conveyed is completely different. Not to mention the concept behind it—a smiling face. To a computer, the characters :-) and :-(are only 33 percent different. But to us, they are 100 percent different.

How do we learn such knowledge, such thought concepts? How do we understand the world? We can see how we *don't* do it by marveling at computer algorithms. Specifically, the most modern algorithms in existence, the "deep neural networks." These are computer systems that are no longer programmed to follow the classic A then B system of logic. Rather they "borrow" from the brain and copy its network structure. The software simulates digital neurons that are able to adapt their points of contact to one another depending on which pieces of data they need to process. Because the cells and their contacts are able to adjust themselves, the system is able to learn over time. For example, if the software needs to be able to identify a penguin, it is presented with hundreds of thousands of random images with a few hundred penguin images included among them.

The program independently identifies the characteristics specific to penguins until it is able to recognize what a penguin might look like.

The advances that have been made in artificial neural networks are huge. Merely by regularly viewing images, such a system is independently capable of identifying animals, objects, or humans in arbitrary pictures. Facial recognition capabilities have even surpassed human ability (Google not only pixels out human faces in its Street View maps but also the faces of cows).[8] But to put it all into perspective: a computer system like this is to the brain what a local amateur athlete is to an Olympic decathlon champion. The comparison is not even the same concept because computers do something that is very different than neurons, in spite of the pithy appropriation of the neurology terms by IT companies who claim they are building "artificial neural networks." In reality, computers are neither replicating real neural networks nor a brain. It is nothing but a marketing trick by computer companies. For a deep learning network to learn to identify a penguin, it must first process thousands of images of one, in a method that follows the maxim "practice makes perfect." But this is not necessarily how the brain works.

Deep understanding

I WAS RECENTLY standing in the hallway with my two-and-a-half-year-old neighbor. He pointed to the ceiling and said, "Smoke detector." I was amazed and had to ask myself what kind of parents did this little boy have. Did they perhaps subject him for weeks and weeks to thousands of pictures of smoke detectors, always repeating the series of images until he was finally able to identify the similarities and characteristics of smoke detectors and to correlate the object? His father is, admittedly, a fireman and so my neighbor already has a certain predisposition toward fire safety tools. But still, had this little human really been bombarded with thousands of pictures of smoke detectors, fire extinguishers, and fire axes that then enabled him to quickly identify the required implement for the next possible crisis? And then did they send him down the hall in my direction once he had finally passed the test with flying colors? No way! That's not how it works. But the question still remains: How was my little neighbor able to identify a smoke detector in a completely new context after only seeing a smoke detector maybe two or three times in his short life?

The answer is that my neighbor did not learn about smoke detectors in the same way that a computer does, rather he understood the idea of smoke detectors. This is something which humans are very good at and which science calls "fast mapping." If, for example, you were to give a three-year-old child never-before-seen artifacts and explain that one very special artifact is named "Koba" or comes from the land of "Koba," the child will remember the Koba object one month later.[9] After only seeing it one time! It gets even better if the child is learning to understand new actions and not only new words. Children who are only two and a half years old require only fifteen minutes of playing with an object before they can transfer its properties to other objects. For example, a child who realizes that they can balance a plastic clip named "Koba" on their arm later realizes that a similar clip, but with a slightly different shape, is also called a "Koba" and can be balanced on one's arm.[10] The whole exchange only takes a few minutes. How would two-year-olds possibly be able to learn an average of ten new words a day if they had to practice each word hundreds of times? No brain has that much time on its hands.

Of course, the brain cannot simply learn something from nothing. From what we currently know, we assume that learning by "fast mapping" allows new information to be rapidly incorporated into existing categories (presumably without even bothering the hippocampus, the memory trainer that you learned about in the previous chapter).[11] But we are even able to create these categories very rapidly—whenever we give ourselves time for some mental digestion. If you present a three-year-old with three variations of a new toy (i.e., a rattle with different colors and surfaces) one right after the other and give each of these the artificial designation of "wug," the child will not easily be able to identify a fourth rattle as a "wug." If, however, the child is allowed half a minute of time between the presentation of each new rattle to play with the item, he or she would then grasp the concept of the wug and be able to identify a new, differently shaped and differently colored rattle as a wug. This seemingly inefficient break, this unrelated waste of time that we would love nothing more than to rationalize away in our productivity-optimized world—this is our strength—if, in fact, we hope to be able to accomplish more than a mindlessly learning computer.

We are very quick to understand categories and are able to grasp the relationship between words, objects, and actions almost immediately. You don't believe me? Do you still think it's only possible to effectively learn something by repetition and practice? Then allow me to give you a counterexample: How long did it take you to understand a newly coined word like "selfie"? A single experience of seeing four posing teenagers snapping a photo of themselves on a smartphone should have been enough.

How quickly were you able to understand the invented word "Brexit"? You probably figured it out fairly quickly. We often understand the world at first glance, but there's more. Once you've understood something, not only can you reproduce it, you can also make something new from it. If Brexit describes the exit of Great Britain from the European Union (EU), what would "Swexit," "Spaxit," or "Itaxit" indicate? Or from the opposite direction, what would "Bremain" or a "Breturn" mean? It's a piece of cake for you to grasp all of the new words because you already understand the fundamental categories of thought. You are able to take these and immediately generate a new piece of knowledge, even if you've never heard of "Spaxit" before in your life!

So much for the topic of frequent repetition and "deep learning." Merely memorizing a bunch of facts is no great art. Understanding them, on the other hand, is. In the future, computers might be able to "learn" about objects and pictures more quickly, but they will never be able to understand them. In order to learn, computers use very basic algorithms to analyze an enormous amount of data. Humans do the opposite. We save much less data but are therefore able to process exceedingly more. Knowing something does not mean having a lot of information. It rather means being able to grasp something with the information in hand. Deep learning is all well and good, but "deep understanding" is better. Computers do not understand what it is that they are recognizing. One interesting indication of this followed from an experiment conducted in 2015. Researchers studied artificial neural networks that had trained themselves to recognize objects (such as screwdrivers, school buses, or guitars). The network was analyzed to find out what, in fact, it had recognized. For example, what would a picture of a robin have to look like in order for the computer program to be able to respond with 100 percent certainty that it was indeed a "robin"? If anyone had expected that a perfect prototype image of a robin would pop out, a sort of "best of" from all the robin images in the world, they would have been disappointed. The resulting image was a total chaos of pixels.[12] No human would be able to identify even a very rudimentary robin in such a pixelated mess. But the computer could, because it recognized the robin only as a graphic representation of pixels and did not understand that it was a living creature. If one taught a computer that Brexit refers to the exit of Great Britain from the EU, the computer would never be able to independently draw the conclusion that Swexit means the Swedes waving goodbye.

Our ability to learn extremely quickly, or we had better say, to understand things, is only possible if we do not "learn" facts and information separately, in a way that is sterile and detached, but rather by creating a category correlation that embeds things and, thereby, leads us to understand them. Computers do exactly the opposite. They are very good at saving data quickly, but they are just as dumb as they were thirty years ago. Only now, they are dumb a little faster. This is because they never take time to reflect on all of the data they have gathered. They don't treat themselves to a break. Computers always work at full blast until they run out (or have their power switched off). But if you never take a break, you cannot ever put the information that you possess to any use, and thus you cannot acquire any knowledge. In order to generate concepts, it is essential to have a stimulus-free space (during sleep). We are able to recognize something at first glance because we don't allow ourselves to be flooded with facts and data but, instead, make ourselves take a break. This may initially seem to be inefficient and perhaps to smack of weakness, but it is actually highly effective. In fact, this is the only way that we are able to comprehend the world, instead of merely memorizing it.

Learning power reloaded

WE SHOULD THUS not treat the brain as though it were an information machine since the most valuable learning processes of the future don't call for us to have flawless memories (that this isn't even possible is touched on in the next chapter), but rather for us to adjust rapidly to new situations. If we start competing with computers, trying to use learning tricks to memorize more facts, telephone numbers, and shopping lists, we are certainly going to lose. Maybe we should let algorithms take over these kinds of basic tasks for us.

Trying to develop the latest learning techniques in order to remember more information isn't what's important. It's much more valuable to improve our ability to think conceptually and to understand. The brain is not a data storage device. It's a knowledge organizer whose major talent can only be actualized once we stop treating it like an imbecile—the way that I did with you at the beginning of the chapter. Sorry about that.

You've now learned the most important ingredients for improving conceptual thinking. Stress is only helpful to learning when it is positive, short term, and surprising. Long-term stress should be minimized by reinterpreting it. Students who are aware of what stress is, for example, have been shown to exhibit better coping techniques in response to stress and are thus less prone to tense up while learning.[13]

When are we best able to learn? When we are excited about it, of course. Facts are not that important. It's the feelings that stick with us. Best is if the feeling is positive. Positive feelings should thus be conveyed at school, university, or in work environments by the teachers, lecturers, or team leaders to promote the best learning. This is much more crucial than the factual content that is taught. My best teacher (my chemistry teacher that I mentioned in the introduction) didn't keep a stockpile of modern PowerPoint presentations on hand, but he was very enthusiastic about his discipline. And when someone is so passionate about the citric acid cycle, then there has to be something to it. That's why I decided to study biochemistry. Not because I found the factual content to be so captivating (that came later), but because I was entranced by his excitement for the topic. It is only when something impacts us emotionally that we never forget it—even if it is a form of positive stress.

Learning is all well and good—but understanding is better. And in order to understand, we need a context. Even small children are able to comprehend the world at an incredible pace if given examples and concrete applications to figure out the "why" of things. This happens, not by dumping data and facts on their heads, but by allowing them to construct meaningful correlations for themselves. If you want students to learn new vocabulary, you could give them a list of words. Or you could encourage the children to come up with a personal story to incorporate the new words. Children will quickly gain the individual context that they need to remember the words. I have long forgotten every word I've ever been given on a vocabulary list. But there are other words that I have only heard once, such as when I lived in California, that I have immediately used and adopted into my vocabulary. At the same time, we should avoid getting drawn into the temptation of trying to compete with computer software and artificial intelligence. When it comes to speed, accuracy, and efficiency, we are definitely going to lose every time. It is much more valuable to remember our human weakness, er, I mean strength. Namely, that we are able, sometimes even at first glance, when we take regular breaks, to happily absorb useless knowledge. Yes, of course it is important that we make good use of computer science and modern technological media in school, as we are going to need these skills to function in the future world. But we shouldn't attempt to think like an algorithm. Subjects such as history, natural science, languages, or philosophy, and a good general, well-rounded education are what empower us to establish ideas and conceptual correlations. If you are digging through Shakespeare's masterpieces and chance across the line, "To be, or not to be? That is the question," you could choose to learn it by heart, copy and paste it as a cool meme on Facebook, or save it onto your flash drive. The latter takes up 42 bytes but means nothing. Alternatively, you could go to the theatre and enjoy watching *Hamlet*—and the phrase will suddenly take on meaning.

In this chapter, we have been able to unpack how to build up much more effective categories of thought. By taking regular small breaks, for instance, one is able to produce a thought concept. Once the concept has been understood, it can then be applied to new situations. Only humans are able to do this. When a "deep learning" computer analyzes millions of images, it will, doubtless, be able to recognize that a chair is most likely an object with four legs, a flat surface, and a backrest. But for us, a chair is not so much an object with a particular shape as it is something to sit on. Once we have understood this, we suddenly see chairs everywhere and can even apply our knowledge to invent, develop, and design new chairs. For example, I recently had one of those bouncy yoga chairs at home. My little neighbor remarked, quite correctly: "A ball!" But when I went over and sat on it, he said: "Oh, chair!" Try teaching that to a computer.

3

MEMORY

Why a False Memory Is Better Than None at All

O N OCTOBER 14, 1994, Tom Rutherford's world collapsed. He was forced to resign from his career as a minister in the Assemblies of God church because his daughter accused him of sexual assault. The twenty-one-year-old daughter credibly asserted that she had been abused multiple times between the ages of seven and fourteen. She claimed that she got pregnant and was forced by her father to have an abortion with a clothes hanger while her mother looked on. Just imagine! Rutherford lost his job, his friends shunned him, and he had to piece his life together with whatever jobs he could find. One year later a new truth was revealed: his daughter was a virgin. Her vivid memories of abuse only surfaced when she had begun visiting a talk therapist for the purpose of stress management. Over the course of more than sixty sessions and through a question and answer game, Beth Rutherford developed a false memory that had never before existed. The situation was not necessarily deliberate, but was undertaken with the best intentions to help the young woman work on past stress factors. However, at some point, she was unable to distinguish her rampant fantasy world from reality. It was only when a gynecological report left no doubt as to her virginity that Tom Rutherford's daughter recanted her statements. She subsequently sued her therapist for a million dollars in order to raise public awareness of the danger of false memories.[1]

I'll admit—the Rutherford situation is an extreme case. However, three-fourths of judicial errors can be traced back to false testimonies.[2] People make claims of having become pregnant under satanic rituals, although they never were pregnant.[3] An accused person can remain behind bars for decades due to witness claims that they were spotted at a murder scene until a DNA test finally shows that someone else was the perpetrator.[4] These scenarios all represent a nightmare that is nonetheless reality, since most of the prosecution witnesses in question were not aware that their memories had gotten out of hand. A classic lying test would not have been able to detect such false memories because, for the witnesses, their statements did not feel false. This is one reason why eyewitness accounts should be handled with utmost caution.

Not only judges, but historians also have their jobs cut out for them in assessing the truth behind the claims of contemporary witnesses. There is still a raging debate about whether or not low-flying fighter planes shot at civilians during the bombing of Dresden, Germany, in 1945. Dozens of Dresdeners claim to remember such scenes from their youth. However, in the firestorm of the bombardment, low-flying maneuvers were hardly feasible over Dresden and were also not recorded in any military reports. The case pits living narratives against dry deployment reports. Who are you going to believe?

Evidently, we are prone to distorting our memories. "That wouldn't happen to me!" you might say. "I am perfectly aware of what happened." But it's not that clear-cut. Take this example: Do you roughly remember the list from the previous chapter? Don't flip back and cheat! Now, I am going to give you four words but only one of them was in the original list. Which was it?

Sleep
Phenoxyethanol
Blueberry
Submarine

Okay. We can rule out submarine. And phenoxyethanol. What about the other two words? Think about it, try to picture the list. Which one do you pick: sleep or blueberry?

You are naturally clever enough to realize that none of the words were in the original list. But most people would spend some time going back and forth between sleep and blueberry, or else decide with absolute certainty that it was blueberry. The word seems to be the best fit.

I'll admit that I knowingly led you astray, as I falsely informed you, firstly by stating that one of the words was in the original list. On top of that, the original list had a few key words (dream, night, strawberry, raspberry) that could cause you to draw associations between the target words "sleep" or "blueberry." And to add insult to injury, the two contrasting words (phenoxyethanol and submarine) made the remaining words seem more plausible. Perhaps these factors influenced you to create a false memory and believe you could remember the word blueberry on the list. But don't kick yourself too hard. In the following pages, we are going to learn why this is not such a bad thing, but is instead another one of the brain's strengths.

The memory chapel

AT THIS POINT, I would like to appeal to your memory of the previous chapter (this time without any ploys or gimmicks). Namely, that information and our brain's memory content are not static but rather malleable components. We don't go around recording our minute-by-minute experience of the world with a video camera and then save the film for the rest of our lives. It is much more likely that we are continuously tinkering with our memory, which is itself a dynamic construct.

Memory is dynamic because it can change quickly. Let me return one last time to the orchestra comparison from the last chapter. In the same way that a piece of music can be varied by how it is played by an orchestra, a piece of information in the brain can also be changed. In addition, an orchestra might have fewer or more musicians playing (or neurons, in the case of the brain), in which case the basic concept of the piece of music (the information) remains the same, but the sound would change.

Our memory is a construct because we do not, strictly speaking, call up a single memory when we remember something. Instead, we recreate it new each time, the way that an orchestra plays the same piece of music again but always somewhat differently than before. Each concert—each recollection—is a unique experience. Once our neurons have stopped activating the corresponding memory, the specific memory is gone though it is still saved in our memory. In an orchestra, this memory would be like the ability of orchestral musicians to listen to each other, to play their instruments at precisely the right moment, and thereby be attuned together. In the brain, information is stored among the neural contact spots so that it can be triggered and recreated again in the future. A memory is thus the ability of the neural network to generate a state of activity (corresponding to a piece of information, a thought, or a memory).

In order for an orchestra to play a new piece of music, or for a neural network to save a new piece of information, three steps must be taken. First, the musicians play the new song for the first time, then they practice it, thereby improving their coordination with each other so that they can finally be ready to play it at a concert. Of course, it is important that the orchestra plays the notes with as few mistakes as possible. And this is where a brain is different from an orchestra. A brain doesn't play the order of notes as they are written on the score but instead alters the melody a little bit each time it practices. The brain's goal is *not* to play (or to activate) something that has been predefined because there is no conductor. It is much more important that the brain plays in such a way that the song flows and that it provides a coherent overall feel while at the same time saving as much cell energy as possible. What this means, however, is that the information changes over time, and the more often it is worked on, the more it will change. Our memory is thus contestable at every step: when it records, when it consolidates, and when it recalls past memories.

The vulnerability of memory

IT IS POSSIBLE for a memory to be altered right from the moment when we are learning and saving a piece of information. This is often due to the fact that information is more easily processed when it can take up as many areas of the brain as possible. In order to better solidify a memory, you will often come up with a few pieces of additional information for the brain. It's like the technique that memory experts recommend, of imagining images or stories to correspond with words or phrases in order not to forget them. Something similar occurs automatically in our brain—and sometimes it goes a little too far.

If you want to research human memory errors under standardized scientific conditions, you somehow have to create these errors artificially. One tried and tested method of choice is a test named after its developers, Deese-Roediger-McDermott, or DRM for short. Test subjects are first shown a selection of terms, which they are asked to memorize quickly. For example, read the following list two or three times thoroughly:

Truck, street, drive, key, garage, SUV, freeway, accelerate, gas station, bus, station wagon, steering wheel, DMV, motor, pass

DRM researchers do this with several lists and then give the test subjects time to think or else distract them with a different focused exercise. In order to do this with you, I will simply continue writing this sentence, adding in an unnecessary clause here, and putting something completely unimportant (irrelevant) into parenthesis over there, all in order to move you a bit further away from the list. Don't look back! Please move on to the top of the next page!

Your task: which of the following word(s) did you see in the previous list?

Steering wheel, car, seat, motorcycle, inspection

Alternatively, I could ask you to write down which words you remember and then compare the lists to see which words match and which words you may have added. By now you have realized that this is a simple variation of the puzzle that I gave you at the start of the chapter. Interestingly, by setting up a carefully devised experiment (one that is not as limited as I necessarily am in the confines of this book), it is possible for researchers to get 80 percent of test subjects to claim that they "recognize" false words, or to put it another way, to facilitate the production of false memories.[5]

The reason behind this weakness of memory lies in the manner in which information is integrated into the brain from the very beginning. When you read a word, you first receive it in the image-processing region of your brain, which is located in your neck area. However, in order to grasp the contents of the word, it must be semantically processed—that is, to relate the new word's meaning to other words with similar meanings. This takes place in the frontal cortex, or more specifically, in the front and side area of your frontal lobe (for those who must know exactly: the ventrolateral prefrontal cortex). Studies show that both true and false memories are generated in an almost identical fashion. Although true memories do tend to show increased activity in the image processing region—since one is closer to the raw information from one's surrounding environment—the further processing stages in the brain are more or less the same for false memories.[6] Or, to put it another way, not only do we perceive a list of words, we shape the perceived truth. We endow it with a meaning and pack the concepts into a mental box.

In chapter 11, we will explore the consequences of excessive inside-the-box thinking or pigeonholing, but for now, allow me to make one remark: invented memories are created in the same way as true memories. Of course, false memories cannot boast the same level of "real" sensory experience, but they are nonetheless integrated into the exact same network. Once this has happened, even if only a single time, it's too late. The brain can no longer distinguish the false memory from the true one. From that moment on, it doesn't make any difference to the brain what is fantasy and what is reality. Or, to quote the neuroscience film classic *The Matrix*, "Your mind makes it real." Regardless of whether or not the experience ever really took place. In principle, our memories live in a dream world of our own creation.

Before we slide down the rabbit hole of a fundamental philosophical-epistemological debate, let's return to the brain's process of memory formation. A process which is not only influenced by our habit of arranging new information in patterns and boxes but also by our emotions and by fellow human beings.

The emotional traps

We are not only prone to making mistakes when trying to recall lists of words; we also err when we need to put them into a social context. At the end of the day, a piece of information by itself is not the important factor. It also matters to whom, what, when, where, why, how it was said. Researchers have studied this by showing test participants various videos. In one video, a person speaks directly to the participants, and indirectly in the other video, by looking off to the side of the camera lens. This has an apparent effect on participants' memories—not of the contents of the words spoken (which they are able to recall well under both circumstances)—but of the conversational situation. Most participants falsely recall having been directly addressed in the video, even when they were shown the video with the person speaking indirectly.[7] While the hippocampus (if you recall, the memory center of the brain) was busy saving the contents correctly, a neighboring region of the brain, the anterior cingulate cortex (ACC), was responsible for the incorrect memories of the conversation and was excessively active in creating the false memories. We seem to take in information subjectively which we also do not save objectively.

To add insult to injury, emotions can lead to memory distortions as well, even in a simple DRM test. For example, if study participants are placed under emotional stress, such as giving a speech in front of an audience or having to take a terrible math test before they are instructed to memorize a list of words, they end up generating a large number of false memories.[8]

However, not all emotions have the effect of misleading our memories. Those types of emotions particularly predisposed to distort our memories tend to have two traits in common: the emotions are intense, and they also appear to match the information that we are supposed to remember. So, if we are in a good mood, we are more prone to falsely remember words from positive vocabulary lists. If we are in a bad mood and stressed out, it's more likely that we will confuse lists of negative words.[9] The best option would be to always be in a good mood when driving a car because this would enable us to be the perfect witness to traffic accidents. Although... research has pretty clearly shown that no one makes for a very good witness to an accident. This is due to another memory-related weakness of the brain.

The perfect memory crime

NOT ONLY DOES our brain have problems saving information, it also struggles to consolidate and retrieve the information from memory since it is susceptible to all kinds of misinformation, which it gladly accepts and adds to the previously existing memory. Our memory is thereby no longer true, though it may seem more coherent.

Just imagine, for example, that you are sauntering down the street when you suddenly hear the sound of squealing tires! You can only guess where the noise is coming from and you turn to look just in time to see two cars crashing into each other. Naturally, you voluntarily offer yourself as a witness to the scene, and this is where the problems start. You only really "half" experienced the crash. You believe you saw how the cars drove into one another, but you aren't quite certain. It all happened so quickly. The brain really dislikes being in a condition of uncertainty (experts call this "cognitive dissonance"), and it is always trying to create a coherent overall picture. If what you perceived is fragmented, the brain will substitute in the rest of the information without you even noticing. Incidentally, this is the same brain that both generates a seamless consciousness and recalls false memories, so you also won't be able to trace the origins of the false memories. In other words, it is the perfect memory crime story—in which the perpetrator (your brain) and the detective (your brain) are one and the same. Both players have very little interest in an explanation, meaning you won't even think twice about the false memory.

Imagination is learning too

OKAY. SO, IF this is true, how should our brain be able to figure out which memories are true? Because the brain does not have any "criteria for truth," it uses a trick in which it only classifies information as real if it activates a large portion of the brain. In other words, if something really *did* happen, it must have left large tracks of activity across the network. This is true in principle since authentic experiences trigger particularly intensive brain activity. If we merely imagine a photograph, the image processing areas of our brain are not as strongly active as they would be if we had the physical photo in our hands and were looking at it. The only problem is that these tracks of imagined memory activity are enlarged retrospectively until they artificially grow to be as big in our memory as real-life stories.

This phenomenon has been studied by showing test participants a series of different photographs from daily life situations.[10] On the following day, the participants were reminded about the photos from the previous day by being told brief descriptions. What they didn't realize, however, was that some of the descriptions were deceptive and falsely described the photos. Misled in this way, some of the participants formed false memories of the original photos and were no longer able to select the correct (original) images from a selection of photos. Some of the participants even claimed to have seen a manipulated version of an earlier photo. At this point, the level of brain activity triggered was very similar for true and false memories, with one decisive difference: the image processing area became more active with the correct memories (since the participants had, after all, really seen these photos). On the other hand, if a participant incorrectly recalled an image, the audio area of the brain was more active (because they mixed up the new deceptive information that they had been given verbally with their actual memory). In other words, if the total amount of activity covers enough area and is integrated into the brain, the memory is accepted as true, even if it is not.

This study significantly illustrates that memories are by no means static but, on the contrary, may be altered retroactively— and that this takes place every single time you bring them out and dust them off. Whenever a memory is in this state (of being hauled out and dusted off), it is particularly vulnerable to external influences. One elegant experiment was able to demonstrate this effect. Participants were first asked to memorize a list of words and then received a new list the next day. Before being given the new list on the second day, half of the participants were asked to try to remember the first list. On the third day, all of the participants were tested on their memories. Some participants who were once again asked to try to remember the very first list got it mixed up with the second list—but only if they had also been asked to remember the first list on the second day. Those who only concentrated on the second list on the second day (and were not asked about the first list) were able to separately recall both the first and second lists.[11] The conclusion: a memory that is in the process of being recalled is in a fragile state and susceptible to corruption by new information.

Peer pressure memory falsification

AS IF IT wasn't enough that we can make mistakes when saving information, as well as altering our memories each time we pull them out, we are also hardly able to defend our memories from being actively manipulated by external factors. Even if we are aware of this, we are powerless and continue to indulge in memory falsification. Peer pressure—the obligation to adapt our memories to those of other people—actively influences our memories.

In order to demonstrate this concretely, participants were shown a two minute documentary film and then asked a series of questions about the video.[12] Directly after viewing the videos, participants made few errors in their responses and were correctly able to recall the details. Four days later, they could still remember the details and didn't allow their memories to be swayed by any false information about the film. This changed, however, when participants were shown fake responses about the film made by other participants. Upon seeing the incorrect answers of others, participants were also drawn toward the wrong answers themselves. Even after they found out that the other answers had been contrived and didn't have anything to do with the documentary, it was too late. The participants were no longer able to distinguish between truth and fiction. They had already modified their memories to fit the group. Interestingly, this peer pressure effect is conveyed through a brain region that neighbors the hippocampus, called the amygdala. This almond-shaped, dice-sized region showed a flurry of activity, particularly whenever the fake responses were shown along with a photograph of the other participants and not merely in writing. A human face increases the feeling of peer pressure and leads our memory further astray toward false conclusions.

It is now possible to list all the ingredients needed to concoct a false memory: an emotional event, a dash of peer pressure, and a habit of frequently recalling a memory, which gives it the opportunity to become further distorted. When this happens, it is nearly impossible to distinguish false memories from true ones. If you would like to cause someone to generate a false memory, it would be best to do it in steps. First, confront the person in question with a distinct (but falsified) memory scenario—for example, that he or she once lost their parents in a store as a small child or that they got into trouble with the police as a teenager. Reinforce this with the fake claim that relatives would be able to back up this situation. Ask your test subject to imagine the event in question and then to think about it for a few days. Then grill them once more with questions, appeal again to their imagination, pressing them for details. Usually by the second sitting, detailed but false memories start to emerge. In this way, it's not only possible to get a twelve-year-old to contrive absurd stories (for example, that he or she was abducted by a UFO[13]), it's also possible to convince 70 percent of adult participants that they had once committed a crime, even if such a claim was completely fabricated.[14]

The start of this chapter showed what can happen if one spends weeks, or even months, performing such imagination exercises for false memories, and this very important point cannot be emphasized enough: don't depend on your memory! It is never one hundred percent correct, and it has more likely than not been embellished, distorted, or partially erased by your own brain. You have been influenced by other people, and you are subsequently unable to tell a false memory from a true one. Even the brain is no longer anatomically able to distinguish one from the other since the activity patterns of true and false memories are nearly identical. There are, however, two small but fine exceptions. First: correct memories trigger more activity in the hippocampus and in the image processing regions (since one experienced the true memories, after all). And second: fake memories result in increased activity in the frontal cortex (presumably because the brain must exert itself somewhat in order to come up with an artificial memory image).[15] However, the neural network is so similarly and expansively activated for both types of memories that it becomes impossible for you to be able to tell the difference anymore. As I mentioned earlier, at this point it no longer even matters whether you know that you have been falsely informed. Once an incorrect memory has fallen down into the well and been absorbed, it becomes as authentic as a true memory. Reality and truth are thus two wholly different things.

Memory rescue

BY NOW YOU are probably wondering what you can possibly do to save your true memory from being taken in by a fake one. In principle, there's not much you can do because this memory system is quite robust and is going to go on leading you around by the nose. However, neuroscience has a few findings that show it's possible, under certain conditions, for our memories to become even more robust.

Possibility 1: You grow older. Specifically, memory improves and people are less prone to develop false memories when, for example, they are warned about the pitfalls of false memories before taking a DRM test. If I were to have you repeat the same test that you took at the start of this chapter, you should have learned by now that you are sometimes going to fall into a habit of mental pigeonholing that you formed while you were saving information. Interestingly, this tendency to be cautious becomes more pronounced the older one gets. This is why it is possible for a sixty-six-year-old person to effectively shield themselves from new false memories if they have been duly warned on that topic before taking a memorization test. Younger people (ranging from eighteen to twenty-three years old) who receive the same warning still fall prey to false memories.[16] Their brains are apparently more eager to go about constructing mental boxes, which serve to obscure the facts. Older brains, by contrast, feature more control mechanisms (or, to state it in a negative sense: they are already stuck in their own ways and are therefore less vulnerable).

Possibility 2: You take birth control pills. Women who use hormonal contraception perform just as poorly on the DRM test as women who are not on birth control. But they are less susceptible to any later misinformation. If you show them photographs from daily life scenarios and then later mention that the scenes appeared differently (for example, that a person is standing in front of a tree instead of a door), they will not incorporate this false information into their memories.[17] The reason is presumably due to the fact that female sex hormones decrease one's receptivity to minor details (especially when they are spoken about rather than seen). To put it another way: a woman on hormonal birth control leaves you not with a poem but with a photographic impression. Dear male readers, please keep this in mind when you decide to offer a gift to the woman of your heart or think that she won't mind if you offer a lyrical description of yourself in place of a dashing photo. And one more warning, before you dope up your next testimony witness with oral contraceptives: whether the same outcome is true for men has never been studied. This is most likely due to a lack of willing male test subjects.

Possibility 3: Be aware of your memory's weaknesses—preferably in the very moment that you are experiencing a new bit of information for the first time and then committing it to memory. Do not underestimate the fact that you often tinker with your memory and constantly distort it. When it comes to recalling something as precisely as possible, it may be harmful to try imagining it so intensely. Often the first (and the most likely unfalsified) memory is the best and most objective one, and witness testimonies can benefit from allowing witnesses to assess their level of certainty right off the bat. If your goal is to retain the original memory, less feedback is more.[18] The more often we compare our memory with the comments, assessments, and perspectives of others, the more we distort it. This may sound awful, but there is an important underlying principle to it.

Why false is sometimes better

AT THIS POINT in the chapter you have learned—if nothing else— just how shoddy your memory is, at least, from the perspective of accuracy. At the same time, our brain has always been capable of saving information with exact precision. But it doesn't. The reason is because a smidgen of false memory can have enormous advantages.

One advantage of this particular memory weakness is obvious: it saves time and mental effort if we don't have to remember all of the details from a list of words or an event but only remember the corresponding context. If you see fifteen words that fit the category "car," you might easily add an extra word from the same car-related category, but you will not add a word having to do with hobby gardening. In other words, it's generally much more important for the brain to recognize the big picture than to hone in on the details. We don't draw detailed information from our surroundings in order to piece it back together again like a puzzle. Instead, we tend to use individual bits of information (words, images, objects) as cues to which we then invent a matching framework of meaning. This allows us to navigate very quickly instead of getting bogged down processing a gigantic pile of details and information from our surroundings. This is why we are able to locate objects much more quickly if they seem to fit their setting[19] (i.e., a pan in the kitchen instead of in the bathroom). It's a mental shortcut of sorts that saves us energy, though unfortunately with a little less precision.

Imagine that you are tasked with understanding a situation by quickly and intuitively piecing together words or objects. For example, cross out two words that don't fit into the following list:

House, tree, bush, cabin, apartment

What has to happen in order for you to be able to extract "tree" and "bush" from the other words? Concentrating on the particulars of each word is not as important as seeing their semantic characteristics (their meaning) in relation to the other objects. It doesn't matter if, three days later, you can't remember whether it was house, cabin, and apartment, or home, cabin, and apartment. The main point is that you still have the concept of "shelter" in mind. Interestingly, the brain region that processes meaning is identical for both true and false memories (the lateral prefrontal cortex, the part of the frontal cortex that is also involved in processing the meaning of words). This may be one reason why people who are particularly prone to forming fake memories tend to perform very well on association tests.[20]

Seen from this angle, it's possible to interpret such over-exuberant false memories somewhat differently, as an especially creative strength of the brain. If our brain were to always function with precision and perfect replicability, like a computer that is able to open up a saved photo with the same quality, we would never have the opportunity to use our memory for new thoughts. Astonishingly, the formation of false memories goes hand in hand with the formation of new ideas and problem solving.[21] Test subjects were particularly spontaneous and intuitive in coming up with umbrella terms for groups of words if they had first been stimulated to create false memories. The ability to think associatively, to draw correlations, or invent them is only possible if we free ourselves from rigid forms of memory and recollection. Errant memories are a necessary by-product of the way in which we think—namely, that we are not so much fixated on data or details as on meaning and stories.

It's not true, but it sure rings true

MEMORIES HAVE TWO main functions for us. We use them to construct an identity with our past and to learn from our experiences in order to improve going forward. For both of these functions, our memories need to be flexible, not static. The caveat is that flexibility also implies vulnerability.

The more we remember, the more we embellish our recollections and thereby distort our memories. However, in order to plan for future events, this is precisely the trait that we need. A *what-would-happen-if* thought experiment only works if we aren't clinging too fiercely to specific details of the past but, instead, let ourselves go a teensy bit mad. This madness occurs spread across a collection of about half a dozen brain regions that are predominantly concentrated in the frontal and parietal lobes (and in the hippocampus network). It doesn't matter too much which specific regions these are. What matters is that these regions are as much involved in simulating future events as in "recalling" things that already happened.[22] In other words, in order to be able to imagine something happening later on, we have to deconstruct that which has already taken place and to creatively glue it back together again, like a collage. Of course, this pushes against our desire for a dependable memory and—admittedly—can possibly lead to a mismatched or falsified memory in the end. However, the advantage is much greater; namely, we are able to imagine virtually any possible future (even one that is impossible). It is only by accepting our memory failures that we are able to have and entertain new ideas.

And if you're worried that we forget a lot of things or remember them falsely, please don't forget that memories are not obligated to explain the world as it is. We use memories much more to help us feel comfortable in the here and now. Studies have shown that people very deliberately (though not necessarily consciously) falsify memories of their past in order to generate a harmonious state in the present. For example, if one asks a group of students to recall their abilities "only a short time ago" at the beginning of the semester, they evaluate themselves as having been similarly capable and experienced as in the present moment when they are being questioned. But if a second group of students is asked to remember how they began their semester "back then, quite a while ago," they estimate their earlier selves as being more naive and immature—even when the start of the semester is no further in the past than it was with the first group of students.[23] The more you were a blithering idiot in the past, the better you appear today. Our past self is a fabulous scapegoat because it can't defend itself. This is how we can excuse the negative and persuade ourselves of the positive, twisting the past in order to construct a consistent image of ourselves.

In general, every one of our memories is false and, each time we recall one, it becomes even more false. But if this wasn't the case, and if our memories were imprinted once and for all at the moment when they occur, we would never be able to go in afterwards to "update" and expand these memories later on. This kind of unmodifiable, static memory prison is not a very nice state to imagine, especially because you would then be too inflexible to imagine much of anything anymore. It is thus a good thing that we often make so many mistakes when we are remembering things. Maybe our memories are not quite as true, but they certainly ring true.

4

BLACKOUT

*Why We Choke under Pressure and the Secret
Formula for Fending Off Stage Fright*

THE YEAR IS 1998. On April 25th, Comdex, one of the largest computer trade shows in the world, is underway in Chicago. Bill Gates is about to premiere his new super operating system, Windows 98, to the world. His colleague, Chris Capossela, plugs a scanner into the PC to demonstrate just how flawlessly the device is able to recognize the new Windows system. But what happens next is not supposed to happen. Zap! The computer crashes. An enormous blue-screen error message is projected onto the screen behind the two men, an embarrassing error clearly visible to every single journalist who is present to cover the event. The hall erupts with laughter, and somewhere in California, Steve Jobs rubs his hands together in delight. Bill

Gates pauses for a moment before quipping: "That must be why we're not shipping Windows 98 yet." In spite of everything, he still manages to crack a joke. No computer in the world would be able to improvise in the same way. After all, mistakes can happen to anyone. In this regard, a computer is only human. And without such innocent mistakes, the world would be much less colorful (or blue, in the case of the Windows error message). You might be interested to know that Chris Capossela was not fired over this incident but continues to work for Microsoft as chief marketing officer—where he now premieres the latest software all by himself. Because anyone who is able to survive a moment like that can surely handle tough times.

Take heart. A mental blackout at the worst possible moment can happen to even the best of us. You can practice and train as much as you're able, but the truth is that under pressure, we all make the worst mistakes. It is an embarrassing truth for sure, but one that is also human. Adele blanks on the lyrics to her song in 2016 in Manchester. Christina Aguilera mixes up the second and fourth lines of "The Star-Spangled Banner" at the 2011 Super Bowl. Italian player Roberto Baggio misses the all-or-nothing penalty kick at the 1994 soccer World Cup against Brazil. These are all examples of what can happen if one relies too much on one's brain. Even a person who is a master in their field is not exempt from such intellectual blunders.

For those of us who are mere mortals, mistakes seem to happen precisely at the moment when we would most like to avoid them. The more pressure we feel, the harder it is to perform flawlessly. This happens not only up on stage in front of thousands of screaming fans, but also when we are taking exams, or being interviewed, or while making a presentation to our colleagues. Our brain turns to mush in the very moment when we need it to do its job well. The brain is reminding us, once again, of its very impractical neural characteristic—namely, that it is not an organ like the heart or liver that functions evenly and regularly and always at the same pace (though a case might be arguable for the latter). No. The brain's performance fluctuates. And sometimes it even fails in its primary task.

But why does this happen? Why does our brain become particularly prone to fault whenever it faces heavy external pressure or stress? And is there a secret formula that we might employ to help us conquer our stage fright, test anxiety, or public-speaking jitters and instead give us a boost so we can put our best foot forward at the critical moment? Because there are people who seem to be exceptional performers—whether they are entertainers or athletes—who appear to shine in the do-or-die moment. How do they do it?

The step-by-step trap

THERE ARE MANY situations that carry the risk of choking under pressure. This is the reason why there is variation in the brain's processes. Some moments are in fact predetermined for us to lose our mental cool and to tense up. Allow me to pause here to list the Top 3 most typical mental misfires that we experience under pressure.

Mental misfire #1: the step-by-step trap. This tends to happen especially often whenever practiced and precise movements or procedures are required. Typical examples are precision sports such as golf, billiards, gymnastics, but also hurdling, ski jumping, or soccer penalty shootouts. Surgeons, musicians, and other artists also regularly follow an automated program that does not allow for mistakes. Their necessary actions are either simple (such as during a soccer penalty shootout) or well practiced (such as the hand movements of a surgeon or concert pianist). One therefore practices these actions extensively in advance in order to avoid error at the moment when it really counts. It takes an estimated 10,000 hours of practice to gain mastery over an individual skill, regardless of how talented someone might be. In other words: if there's one thing all skilled people have above anything else, it's that they have put in a lot of time.

Through training and practice, one is able to automate a sequence or motion. After this, the motion is no longer located in the cerebrum's conscious but rather in our cerebellum's subconscious, where our inner autopilot is seated, so to speak. But even when we have managed to almost perfectly master an action, we still find that we are unable at times to call it up when we need to. This is because we have a tendency to become especially attentive under pressure in order to avoid making mistakes. Conscious attention, however, is processed in our cerebrum, and this area of our brain works a lot slower and more inefficiently than our motion-optimized cerebellum. Thus, at the most inopportune moment, we start concentrating on the concrete process that our action requires, causing our clumsy cerebrum to get in the way of our pre-automated and efficient cerebellum. The result is that, instead of allowing our polished movements to simply flow, we suddenly think about every single step, which in turn causes us to lose our fluidity of motion. It's not really that hard to kick a penalty shot into the net. But when it comes down to the crucial moment of performance, even the easiest task becomes a near-impossible feat. Just ask the English soccer team, which has one of the worst penalty-kick records in international soccer.

Whatever you do, don't picture
a red plush rabbit!

SIMILAR TO THE English soccer team's tendency to choke when faced with an all-or-nothing penalty kick, the Italian player Roberto Baggio's brain wasn't working all that badly during the high-pressure situation in the 1994 FIFA match against Brazil. In some ways, his brain was working too well. Why? Let's zoom in a bit closer on that penalty kick moment. The player is standing concentrated and ready to shoot in front of the penalty area. His only task is to get the ball past the goalie and into the net. And not to hit the post or go outside of it. And in that moment, in the very second that he doesn't want to, he kicks it straight at the net's crossbar. A phenomenon that neuropsychology calls "ironic effect on performance." One ends up doing exactly the thing that one wanted to avoid.

The cause of the missed shot can be attributed to the fact that our brain has two systems of action: an operative and an observant system. The operative system is responsible for planning and carrying out all possible movements for an operation (i.e., during a penalty kick, calculating each step and ensuring that the foot is turned at the correct angle in the correct moment). Meanwhile, the observant system is scanning external conditions to recognize any problems, such as a crossbar that one might accidentally hit with the ball. In this case, it informs the operative system that it should adjust its movements accordingly to take this potential problem into account. So far, so good. If everything always went as smoothly as this, no one would ever choke up under pressure.

But the working capacity of the operative system is limited. Especially under pressure, some of its thought resources are exhausted by feelings of stress or anxiety. The observant system, however, continues to run and to offer up threatening loser-scenarios to the brain's consciousness. The result is that our focus on precisely the mistake we wish to avoid becomes stronger and stronger and eventually overwhelms our operative system. The operative system can no longer defend itself because too many resources have been taken over by our anxiety. You end up kicking the ball straight at the crossbar because you are thinking of not kicking the ball at the crossbar. Interestingly, this tends to happen most to those players who have neurotic behaviors[1] or who try their best to veil their insecurity by acting cool. But this only makes matters worse, since such an act of coolness only uses up more thought resources from the cerebrum. Maybe this is one explanation as to why Argentinian soccer superstar Lionel Messi's penalty-kick record is less than perfect. But that's just a guess.

The more one concentrates on the step-by-step process, the easier it is for us to focus on the one thing we don't want to do. It's similar to what happens when you picture a red plush rabbit right when I tell you not to do so. When you read the above subtitle, your brain's observant system generated the message of the red plush rabbit so that your operative system would know what it is supposed to suppress. Perhaps this worked well at the beginning, and you instead pictured a yellow squeaky duck before continuing to read. But by processing each of the written words here step by step, your operative system is left with less and less capacity until it finally capitulates to the constant warning of the observant system ("Whatever you do, don't picture that silly rabbit!") and suddenly the picture of a red plush rabbit springs into your consciousness.

What helps is to turn your attention away from the concrete threat. A little bit of distraction at the right time can help the overactive observant system to relax. If this is done with little warning, the operative system doesn't have time to go in the wrong direction either. However, this only applies to simple and automated types of activities—for example, golf. The basic movement of golf—though I don't mean to offend any readers who might be golf fans—is relatively simple. Nevertheless, when one is under pressure while putting, it's possible to miss the shot even at close range. But if experienced golfers suddenly stop paying attention to the action of tapping the ball, but instead focus on playing faster, they tend to hit better than when they are concentrated on hitting the ball into the hole.[2] The result is the same when golfers are told to listen for a particular tone while they are putting. This small amount of distraction reduces mistakes, as long as the physical action is well rehearsed and can be done on autopilot. When our cerebrum is suddenly given a new task, it can no longer interfere with our cerebellum.

If you find yourself concentrating too much and thereby tensing up under pressure, it might therefore be wise to try distracting yourself a little with something else. Pause and look out the window briefly, let your thoughts drift to something else, recall a pleasant memory, play it through and linger for a few seconds and then, as you turn back, don't concentrate on your task deliberately but simply act. Just as my track coach always said to me: "Henning, you think too much." Such a criticism is offered far too seldom in the modern world.

The distraction trap

Mental misfire #2: the distraction trap. This typically occurs whenever we are required to carry out a sophisticated mental performance during an exam or in an interview. While it's best in a penalty kick or golf tournament to avoid thinking directly about the step-by-step process and to allow ourselves to carry it out automatically, an exam is a totally different situation. In this case, conscious thought can actually help, and requires focus rather than distraction. Someone who loses their concentration during an exam risks sacrificing mental strength to unbeneficial nonsense thoughts. For example, you start thinking about the consequences of the exam, pondering everything that could go wrong, or wondering what type of impression you might make on your potential future supervisor during an important job interview. Or, even worse, you get scared of saying something wrong and in that very moment you really start to freeze up.

As we have just learned, our brain does not have unlimited mental reserves for a task. The more complicated the problem, the more we require the frontal region of the cerebrum (the so-called prefrontal cortex) to find the solution. The only thing is—the reckoning capacity of this brain region is limited and distracting thoughts eat up its mental resources.

The most damaging type of thought we can have in an exam-like situation is the fear of failing. Interestingly, this kind of stage fright or its educational variation—test anxiety—tends to be the Achilles heel of our brain, particularly for people who have mastered the required task! For example, studies of test subjects during mathematics exams have shown that people who usually tend to be very good at math perform especially poorly under pressure.[3] If the test subjects are required to complete a problem within a certain time limit or if their abilities are evaluated by strangers, some of the people who are most capable end up falling behind others who are less capable at calculating math problems. The explanation for this is that fear, or anxiety, works like a magnet in the brain, attracting and stimulating the regions that are also responsible for pain sensation,[4] and uses up essential thought resources. Math whizzes get so worried about buckling under pressure that they no longer have enough mental capacity available for their calculations. So, what do they try to do? They try to save mental capacity by using less stress-prone shortcuts and rough calculations (techniques which the less gifted test takers have been using from the start). Although these simplifications are easy to use under stress conditions, they aren't very precise. Thus, the better mathematicians perform worse under pressure than their inferior counterparts, who have already reached their limited mental capacities. This also means, though, that test anxiety cannot be combatted by additional practice. Whoever has the capacity to do more also loses more during an exam due to their anxiety.

A much better solution is to combat the anxiety directly by simulating the pressure situation in practice, thus growing accustomed to it. One can do this with so-called pressure training (*Prognosetraining* in German), in which an imitation competition or exam scenario is practiced as preparation. This allows you to grow accustomed to the pressure. For example, if you are preparing for an interview, you might try practicing by speaking with someone else who allows you only a single opportunity to answer each question. If you get mixed up or muddled in pressure training, you are not given a chance to correct yourself. Just as it is essential in this type of preparation that you are only given one chance to perform well, it is equally important that you practice under the watchful eye of another person. This is because the feeling of being observed by someone else is extremely distracting and is like poison for our mental capacity. Fortunately, an artificially simulated high-pressure scenario offers us a safe space to practice experiencing this feeling in preparation for the actual performance situation.

That's how the English soccer team broke its penalty kick curse during the World Cup in Russia in 2018. No national team has a worse track record when it comes to penalty kicks, partly because no one ever really paid attention to how to fight the competition anxiety of an all-or-nothing shootout. Gareth Southgate, himself the embodiment of poor English penalty shooting after he missed the deciding kick in the European semifinals of 1996, decided to change that. Learning from his more than twenty years of failure, he turned his experience into the best remedy: as the national coach, he initiated precise pressure training before the 2018 World Cup. During penalty-kick training, he instructed his team members to heckle and taunt their opponents. To make the experience even more like the real game, the players practiced penalty shootouts at the end of exhausting training sessions. Furthermore, every step of a penalty shootout was prepared for in advance. The team analyzed the practiced rituals of American football teams at the Super Bowl—where every team member must learn his individual role on the field. No improvising, less thinking, just doing. They practiced the walk from the halfway line to the penalty-kick position. After every shot, the goalkeeper grabbed the ball and handed it over to his teammates to regain control over the situation as quickly as possible. They set up a list of the penalty kickers months in advance to avoid the faulty volunteering process in the heat of the moment. Or as Southgate put it: "Making sure that there is calmness, that we own the process." No wonder England penalty-kicked out a win over Colombia in the round of sixteen in Russia. Congrats, I say as a German soccer fan. Redemption after twenty-two years, this is how winning is done.

Control is good. Trust is better.

IN ORDER FOR our brain to give its best possible performance under pressure, it has to concentrate concretely on the task at hand. In some situations, it can be helpful to shift focus briefly to something else in order to allow our automatic motions to kick in (i.e., a penalty kick). But as soon as the distraction is not deliberate but is rather prompted by external factors, the brain begins to lose the energy needed for the task. Anyone who has ever had to parallel park knows what I mean. When no one else is looking on, it's a piece of cake, and you can slip your car perfectly into the open space with no problem. But if ten teenagers happen to be standing on the nearby sidewalk, jeering and filming your efforts with their cell phones, even the most advanced parking software isn't going to help you.

Performing under observation throws us off, even if we don't want it to. Even professional pianists press down measurably harder on the keys in front of an audience than when they are playing alone (and they don't even notice they are doing it).[5] And if we are instructed to continue squeezing an empty water bottle at a steady pressure, we inadvertently squeeze it more tightly when we feel we are being observed.[6] Our nerves apparently get flustered under pressure situations, and we become much less precise in our actions. Thanks to neuroscience, we now know exactly which nerves to blame. These nerves are located in a region of the brain just above our ears (inquisitive minds might like to know where—in our lower temporal lobe), an area that is normally involved in controlling our actions. If we are under observation by others, this region is deactivated by a neighboring part of the brain. It's as if there was a deliberate suppression of our most precise control mechanisms whenever other people are watching us. This is bad luck—and what's more, it seems that the very mechanism that causes us to mess up in front of an audience is hardwired into our brains. What can we do?

At least we know for certain that there is one thing we should *not* do with respect to other people: try to control too much. In our day and age, it seems everything has to be regulated and every step monitored so the process might be optimized later on. This might work well with machines, but not with human beings. Because if you try to control humans, they end up making more mistakes than usual. Trust is good, control is better? Forget about it! Neurobiology very clearly shows that those who try to control other people end up losing a lot in the process. First you lose their trust, and then you lose their ability to perform. Of course, it's important to evaluate whether or how well someone has performed their job, but it's best to save this until the end of the process. Trust your fellow humans up front and clearly convey to them that the goal is about the end product, not about their method of arriving at it. Someone who is constantly looking over other people's shoulders is ultimately doing more damage than good.

The overexcitement trap

Mental misfire #3: the overexcitement trap. This most often happens when we are going to speak, or lecture, or appear in general in front of an audience. If we compare the various typical situations in which we buckle under pressure, we can see they all have one thing in common. Regardless of whether we are giving a presentation, interviewing for a position, or taking a final exam, we always have a lot to win or lose. This prospect of punishment or reward stirs us up to an unhealthy level of excitement that keeps our brain from functioning properly. Peak performance can only occur in a very narrow corridor: too little pressure and we perform just as poorly as when the pressure increases tenfold.

A psychologist might be satisfied with this explanation. Psychologists see the brain as a black box. The brain receives an input from outside (too much excitement) and sends an output back (poor performance). But for neuroscientists, such an explanation is obviously too simple. We want to know *what* goes on inside the black box, the brain, when it becomes too excited. By doing such research, what we have found is that it's not always good for us to be paid for our work. Because once we start to think about rewards, our rate of error also increases.

The reward paradox

IMAGINE YOU ARE playing a simple computer game: Pac-Man, for example, in which you are moving a figure through a two-dimensional labyrinth while collecting objects. In order to make your task more challenging, these objects run away from you

83

and award you with varying numbers of points whenever you are able to catch them. It is a very simple game that any twelve-year-old would quickly tire of before turning to share their more exciting Snapchat photos. This is why it's only possible to get people to play such games in the lab by offering them a significant reward. A British study offered five pounds (about US$6.50) whenever participants managed to catch valuable objects in the labyrinth, and for the less valuable objects, participants only received fifty pence (about sixty-five cents). Intriguingly, the experiment revealed that the participants were less successful in catching valuable objects, even when these objects moved in exactly the same way as the less valuable objects. Players were able to gather the fifty-pence objects without any problems, but whenever they started to pursue the five-pound objects, they would make mistakes more frequently and veer off in the wrong direction. Merely the prospect of a large reward led to increased mistakes and decreased performance.[7] Perhaps we should share these findings with bankers the next time they insist on claiming their bonuses?

Brain scanners, unsurprisingly, showed that the reward regions of the brain were particularly activated whenever participants got close to the valuable objects. Reward is processed in the midbrain region, which, in this case, works as a sort of antagonist to the attentive cerebrum. The more the reward region was activated, the less activity was recorded in the cerebrum, which controls our actions. This means that we get so excited at the sheer prospect of a reward that we are no longer able to control our actions with any precision. In other words, if you constantly hold a carrot up to a horse's nose, it is going to run faster, but it will also step into potholes much more frequently because it no longer sees them. It's hard to reach your goal when your horse has a broken leg, no matter how big your carrot is.

By the way, the same thing happens when the tables are turned, and we are pondering possible punishment. In such cases, we end up performing just as poorly as we do when faced with a reward. This is because the control mechanisms responsible for "negative reward" (or punishment) take place in the same brain region as positive rewards and therefore have the same effect. One of the most violent forms of punishment is social rejection. This is why many people are afraid of speaking in front of a lot of people. In surveys, the fear of public speaking surpasses even other widespread fears, such as the fear of heights or of spiders. It's no wonder. If you are high up on stage looking out at an auditorium, it can look pretty scary and, when full of people, contains even more legs and eyes than a spider.

Conquering the fear of public speaking

WHEN YOU ARE preparing to deliver a speech or lecture, all of the bad things that we have just discussed come together: under pressure, we focus too much on every single step, thereby making ourselves vulnerable to ironic effects on our performance (doing the one thing we most wish to avoid). Being under the watchful eyes of audience members saps us of essential mental resources needed by our cerebrum, thereby suppressing rehearsed patterns of action. And as if that wasn't enough, there seems to be a lot at stake so that the reward center of our brain becomes overly excited, draining the juice from our conscious cerebrum.

Help! What can we do? Firstly, it's important to clarify that pressure and any resulting stress reaction are not necessarily a bad thing. The brain has its reasons for developing mechanisms to alter our performance under pressure. The brain has to be able to get the most out of its thinking abilities. Athletes, artists, or managers who are totally relaxed prior to their event or important decision will never reach peak performance. A certain amount of healthy nervousness is imperative for us to give our best.

When we suddenly feel ourselves wanting to think through every single step under pressure, this is, in fact, a mechanism that functions to help us increase our precision. When our brain starts to wander to disconnected thoughts under pressure, the purpose is to help us uncover new, spontaneous solutions for a problem even though it is simultaneously breeding fear and distraction. And whenever the prospect of a reward starts to make us feel jittery, it is only because this has given us a boost of motivation. It's only when these brain behaviors become exaggerated that they start to go in reverse and make us weaker. In a high-pressure situation, we start to focus too much thought on our problematic state with the result that it becomes worse. In such cases, less thought is sometimes more.

In order to build resilience under stressful conditions, it is imperative for us to correct the misconception that our physical reaction is a bad thing. In psychology, this is called "reframing" a situation. When you explain to test subjects that sweaty palms and a pounding heart are a good thing for helping us to achieve our best possible performance, they perform better on cognitive tests than when you simply leave them alone with their stress.[8] Nor is it lying to tell them this, because stress actually *does* help us to be more efficient. As long as you are still able to hold your pen, sweaty palms are not too much of a problem.

You cannot simply turn stress off, but you can train your brain to handle it better by recognizing weaknesses. In the case of exams or public appearances, it can help to practice pressure training. If you are rehearsing a speech, you may find yourself frequently making the mistake of starting over from the beginning every time you misspeak. This method is wrong. It is precisely this kind of derailed thought that you want to avoid during a speech. If you tend to do this in rehearsals, you are actively training yourself to blank out during the actual speech. A much better idea is to practice giving your speech under pressure. Give yourself only one single chance to recite your speech error-free. (Although "error-free" can also mean simply carrying on without stopping, even when you mess up. Rather than getting caught in your mistake, you can spontaneously embark on a new rhetorical path.) This practice helps you to develop specific defense strategies for facing possible mental misfires.

The more rigid and inflexible your plan for your speech, performance, or other type of high-pressure situation, the more likely it is that you will deviate from your plan, thereby making a mistake. In preparing for a speech or lecture, you could, of course, learn the words by heart, but in that case, you would also have to recite it perfectly. Every single moment of uncertainty could lead to a potential fiasco and every forgotten clause to a possible mental blackout. Just like a penalty kick: if you are too concentrated on the crossbar, you are going to kick the ball straight at it. If you are aware of the exact wording of the lecture because you have memorized it, you will also be aware of exactly where you've messed up. It's a much better strategy to set a rough direction for yourself. What is the main message, the image, the statement that I wish to convey? What do I want my listeners to take away? Of course, you can still prepare yourself in more detail for complicated passages and practice running through them multiple times. But a lecture only really comes to life once it becomes something more than the error-free version that you practiced a dozen times previously.

So, what can you do to get over your stage fright or fear of public speaking? Anyone who harbors the fear of failing in front of an audience should try to visualize the pressure situation as intensively as possible before it takes place. Studies show that it helps to describe in detail the sequence of a speech or test scenario.[9] You should then play out various possible scenarios in your mind in order to break down your fear of them. What could possibly happen? Choking under pressure is not a sign of weakness—but of humanness. The same traits that can lead to a mental blackout are also those which animate your speech: your emotions. It is much more interesting to listen to a passionate, though perhaps somewhat shaky speech, than a perfect, squeaky-clean address that might just as well have been delivered by an emotionless lecturing robot.

One more thing: there is no such thing as an unprofessional glitch; there is only an unprofessional manner of dealing with it. Don't sweep your mistakes under the carpet. A lot of people try to hide or mask a mistake as quickly as possible when they are in front of an audience. In doing so, they often get so occupied by their cover-up operation that they lose the thread of their presentation. Instead, admit your mistake shortly and sweetly, thereby dismissing it completely. It's better to prepare an alternative plan, a sheet with key words, for example, that you can return to and regain your footing. If something goes wrong at some point, you won't suddenly find yourself like a deer in the headlights but will rather be able to pull something good out of your sleeve. Of course, this implies that you have already placed something up your sleeve beforehand. But never forget: being prepared is one thing. Embracing the blackout is another. Or, as Roberto Baggio put it after missing what is possibly the most important penalty kick in soccer history: "The only people who miss a penalty kick are those with the courage to take the shot."

5

TIME

—

Why We Always Misjudge It—Thereby
Forging Valuable Memories

C AN YOU REMEMBER when you were younger, and summer
vacation seemed to stretch on endlessly? When you were
always going somewhere you'd never been before and
trying out new things? When you spent hours in the afternoons
hanging out with friends or playing sports or just sitting around
like a couch potato? What a delectable, stress-free time of life it
was back then. And today? Time seems to hurtle past. It feels as
though it speeds up the older one gets. Back when you would go
ambling through the woods with your kindergarten buddy, an
hour would seem like an eternity. Whereas if you take an hour-
long lunch break these days, you know just how fleeting sixty
minutes are before it's suddenly time to get back to work.

Time makes trouble for our brains—and that's why we have trouble with time. Tasks that we thought could be simply whipped out in no time end up being time sinks. Deadlines approach ever more rapidly. Constantly getting bogged down at work or at home, we dash from appointment to appointment, arrive too late to our meetings, or postpone our projects. Time is short. It's a precious commodity. It's no wonder that, when asked, most Germans say they would prefer more time for themselves and their loved ones over having higher financial security.[1]

This is a paradox because we are living in an era where we have more support than ever before to relieve us of time-consuming work, and yet we seem to have less and less time to spare. If you wanted to buy a train ticket in the past, you had to wait in line at the ticket counter and tell the railway official your specific dates. The official would then leaf through a large catalog looking for the best train connections. This was normal and no one complained about it. Nowadays, all I have to do is open up a smartphone train app to download a ticket three minutes before the train leaves, and then I get annoyed if the train is five minutes late. A highly efficient, down-to-the-second train schedule is a good thing, but it also means it is much easier for us to notice exactly how late the train is running.

What's going on is this: making plans might appear to work well on paper, but the brain doesn't really go along with us. It is too at odds with the most basic parameters of any project-planning task—namely, time itself. The brain is unable to measure, adhere to, or even quite grasp time. Time is an artificial construct, a crutch devised by humans in an attempt to organize the world. For the brain, measurable units of time don't exist at all. So how is the brain supposed to handle such a fictitiously contrived arrangement of time, featuring years and seconds? It is not built to recognize minutes or hours, which don't occur in nature. If we decided to divide our day into fourteen hours, with each hour made up of thirty-four minutes and each minute of eighty-three seconds, it would be all the same to the brain. It doesn't matter in the least to the brain how we choose to artificially arrange our time. Our brain is always going to be bad at judging it.

"Time is relative," said the physicist. This is more than a neurobiologist can say, since measurable time doesn't exist for the brain. It's obvious that time seems "relative" to us. Anyone who has ever had to wait for a bus or train for ten minutes knows that those ten minutes drag on like an eternity. Just as anyone who has ever sat across from a new love interest on a first date knows how ten minutes can pass in a flash. But when you later ask the same person which scenario lasted longer—waiting for the bus or the first date—they will be able to recall with intensity the "endless conversation by romantic candlelight," whereas the boring period of waiting at the bus stop is reduced in memory to a mere blip. It's rather strange, isn't it, how the brain deals with time?

Why is that? Why do we try to force ourselves to be good at judging the time? At the end of the day, it comes down to the consequences that we face by not keeping our deadlines or agreed-upon time commitments. Is there anything that we can do to regain some of that time for ourselves?

The planning trap

WE SEE IT again and again. An inaccurate estimation of time can have far-reaching, costly consequences when it comes to planning large projects. I'll avoid taking a cheap and easy shot at the disastrous time management of officials who planned the "new," state-of-the-art Berlin Brandenburg International Airport (origi-nally slated to open in 2012, new projections have now pushed the opening of the airport back to 2020 or 2021). But there is no shortage of other examples of construction projects derailed first by their schedules and then subsequently by their budgets (from the Suez Canal to the Sydney Opera House).

Humans are particularly bad at planning future timelines. The most typical time-wasting mistake at work is what scientists call a "planning fallacy."[2] At least one of the reasons for this mis-take can be attributed to the brain's inability to measure time reliably. When facing a task, we more often than not estimate too little time for its completion. If you've ever had to go around buying Christmas gifts, you know just what I'm talking about. You have a general idea of what you should buy, and in addition you have a fixed deadline of sorts (granted it's a bit Scrooge-like to call a Christmas Eve party a "deadline"). And yet, in the days leading up to Christmas Eve, all panic seems to break loose as hectic buyers scoop up last-minute purchases. On average—and this has been scientifically studied—people finish their Christ-mas shopping four days later than they thought they would at the start of December.[3]

The reason seems to be fairly obvious. We're simply too opti-mistic and don't consider that anything could go wrong during our projects. But the solution is not quite that simple. Because if you ask a person to describe not only an optimistic projection, but also the worst-case scenario, they are just as bad at estimat-ing the time required in both cases. As a matter of fact, the fault lies in our perception of time. Whenever we are supposed to imagine the end of a project, we orient ourselves based on our past experiences. We reflect on how long a similar task took us earlier and use this to chart out our predictions for the future. What we forget, however, is that our memory is completely muddled when it comes to time. Task timelines always appear much shorter in retrospect (especially if the work was monotonous). It's hard to make solid forecasts based on our compressed memory of time.

Here are two hints that might make time more manageable. Firstly, ask someone who is *not* familiar with your situation to give an estimate of how long a task will take. Okay, Christmas shopping may be hard because so many people do it, but when it comes to your work, you can certainly find someone who doesn't know anything about what you do. Studies have shown that professionals are the worst at estimating the processes at which they are experts, even though they are able to describe the processes reproducibly and precisely. The more experienced one is, the more these experiences tend to shrink together in retrospect in one's memory. Therefore, someone who is particularly good at knowing certain processes is, for the same reason, particularly bad at assessing the time it will take them to do familiar tasks.[4] Expertise is not always a good thing.

Secondly, if you frequently carry out similar tasks, keep a written record of how long it usually takes you and use your collection of "end times" as a reference for planning in the future. Sometimes it can be very useful to review the actual measured times and to see, often with surprise, how much longer they took than you seem to remember. At the end of the day, our memory reduces everything to a few blinks of an eye. It's not enough to try to remember past experience because we cannot remember time. It simply does not exist for our brain.

No sense of time

IT'S TRUE. WE don't have any sense of time—literally. Our sensory organs help us to record every other kind of external (and internal) stimulus. Of course, this doesn't always happen without a hitch, as we know from countless acoustic and optical illusions, but it is at least possible in these cases for our brain to measure something real and concrete (a color, a tone, or a temperature). But no such sensory organ exists for time. Rather, our perception of time is configured artificially in retrospect. We experience a sequence of stimuli and then embed these within a temporal construct. In other words, we don't measure time; we "fabricate" it afterwards to match up with our perception.

This would make any physicist scratch his or her head. The brain has an unbelievably awful methodology for coordinating chronological sequences, not to mention that it's extremely error prone. When it comes to our technical world, it's a different story. The average wristwatch is off by about a second per day. Atomic clocks are much more accurate. The most precise models will only gain or lose one second every 140 billion years. Don't ask me why we need to be so precise when the universe is barely even fourteen billion years old. But that's just the way engineers are. Even if they don't know the exact date of the Big Bang, they are still able to measure time down to a fraction of a second.

But the brain's clock ticks differently. Or, to be more precise, it doesn't tick at all. We can't even reliably measure the roughest version of our clock, the daily rhythm. If you are lucky enough to be subjected to tests by neuropsychologists and locked in a room that is cut off from daylight and the outside world, your inner sleep-awake rhythm will adapt to a twenty-five-hour cycle. We can't even get ourselves to conform internally to a twenty-four-hour day! We call this built-in daily rhythm a "circadian" rhythm (that is, "circa one day"). In order for us to become tired when it gets dark, this central twenty-five-hour clock is adjusted by light stimulation to the eyes. Now you can get a feeling for what precision means in neurobiology. So while you have to adjust your watch once a year in order not to be five minutes late, your brain has to adjust one hour per day. If it didn't, our "inner clock" would be two weeks off every year. In which case, you would almost certainly miss every one of your appointments.

Time error "live"

SPEAKING OF WATCHES, are you wearing a wristwatch or do you have a clock somewhere that shows the seconds? Take a look at it for a moment. Do you notice anything right when you look at the clock? Sometimes the second hand (or the blinking digital numbers) seems to last longer for the first second than subsequent ones. When you first turn to look at the clock, the second hand "freezes briefly" and then continues to tick at its normal pace. This is a phenomenon called *chronostasis* (Greek for "standing time"). It is responsible for altering our attention to synchronous movement since we adapt our perception of time to each task.

This works particularly well in facial recognition. The more expressive faces are, the more intensely and longer we seem to perceive them, even if under controlled conditions these faces are shown for the same amount of time as neutrally expressive faces (such as passport photos). For example, when female test subjects are shown images of female faces that have been classified as unattractive, they study these faces for a shorter amount of time than they do faces which are neutrally expressive or attractive—both of which the test subjects look at for the same amount of time.[5] This is perhaps not surprising, because who wants to spend time looking at an ugly face? But this particular study lacked a control experiment. I am quite certain that heterosexual male test subjects would have spent more time looking at attractive female faces than neutral passport photographs...

In any case, this serves to show us that our perception of time is significantly influenced by our surroundings. A cheerful face is all that's needed to end our objective perception of time, but only if it's possible for us to imitate the facial expression (which is usually what happens whenever we look at a photo). When test subjects are asked to guess the amount of time they spent looking at cheerful faces, they estimate more time for facial expressions that they can imitate. If they happen to have a pen in their mouths, which prohibits their mouth muscles from imitating the photographic expressions, the effect vanishes.[6] So, if you would like to prevent yourself from getting distracted by attractive faces in the future, you now know what to do: bite on a ballpoint pen and you'll no longer have problems correctly estimating the time. This just goes to show you that neuroscience can have practical solutions to everyday problems.

All joking aside, our perception of time is subject to naturally occurring biological fluctuations. Interestingly, our perception in the present moment is almost never influenced—it only becomes distorted in retrospect. During particularly emotional experiences, one often has the feeling that everything seems to be "running in slow motion," as if the brain was processing the incoming information even faster in the moment. Of course, this does not happen, since the brain always processes things at the same speed. However, if one is remembering the experience, the memory appears to move in slow motion. This phenomenon was studied in volunteers who were dropped in a free fall one hundred feet down into a safety net (using an amusement park for this particular scientific experiment). Researchers expected that the stress conditions of the fall would influence the time perceptions of the participants, causing the fall to seem longer than it was. This was indeed so, but only in retrospect. During the fall, participants' brains functioned at the same pace as usual. The optical processing speed of the brain can, in fact, be measured. We perceive everything that we can see within a maximum of eight hundredths of a second to be the same amount of time. If test subjects are shown a figure with missing elements

Fr xml ti tx

and then another figure with the missing elements

o eape, hs et.

they will perceive both of these figures as a single unit if both are flashed intermittently at an interval of less than an eight hundredth of a second.

For example, this text.

The random characters seem to meld into one whole figure, or sentence. But if the brain had to work faster under stress conditions, it would have to be able to keep both figures separate when switching back and forth between them. However, subjects who experienced the free fall perceived only a single melded figure as did test subjects who did not jump from anywhere (participants were wearing an alternating blinking wrist device during their fall). This shows that the brain functions just as well during a free fall through the air as it does on solid ground. But, falling participants estimated their own falling time as having lasted 36 percent longer than others who they watched falling.[7]

Note: it is possible for the brain to invent a suitable perception of time in retrospect. This is important, because this is what allows us to store exciting experiences particularly well and to perceive almost simultaneous occurrences as a single unit. After all, it is beneficial when we experience our surroundings as a whole, in spite of split-second time delays.

The dubbed voice delay

WHEN IT COMES to bringing our own body movements into harmony with the outside world, our brain has a problem. Our sensory organs process information at different speeds and yet still have to gather everything under a single umbrella. For this, the brain has a trick. It speeds up or slows down our sensory organs, thereby bringing them into synchronicity.

In order to do this, it always orients its perception of time to the underlying goal of an action. In this way, it is a very egotistical organ, since it seems to think that nothing is more important than its own activities. For example, test subjects who are told to push a button to generate a tone perceive the time between pushing the button and the tone as shorter than it actually is.[8] In this way, the brain makes sure that our actions are processed with speed and intensity. But sometimes it does this too quickly. When test subjects are first allowed to get used to the time difference between the action of pushing the button and the ensuing tone, a sudden change that causes the tone to occur exactly at the moment when the button is pushed results in the participants reporting that the tone occurs *before* they have even pushed the button.[9] Because one has become used to the anticipated result of one's actions, the amount of time between each task shrinks. This offers just one more reason why we get bogged down in Christmas shopping.

I sometimes experience temporal illusions when watching foreign films that are poorly dubbed. Whenever films are dubbed into a different language than the original, the actors' lip movements no longer correspond to their spoken words, and the brain can't reconcile the movements in time. However, studies have shown that after only a few syllables, the brain gets used to the dubbing, and we no longer notice that the film does not match the sound.[10] Problems arise, however, when there are language shifts in the film. For example, if one actor is speaking the original language and the other is dubbed, this can give the impression of a linguistic delay. The mouth movements of one actor match the spoken words but not those of the other actor. Our brain does not like this dissonance at all. That is probably why the German film industry is so über-precise in synchronizing movies from their original language into German. Sometimes, the dubbed German voice is even more popular than the original actor or has contributed to making that actor a success in Germany. Will Smith or Samuel L. Jackson: you should probably thank your German voice actors. They have added real value to your acting as well as some true auditory style in German. One reason for your success in

Germany might well be chalked up to the fact that German brains are so bad at recognizing that your lines have been dubbed (precisely because we perceive time so badly). To be fair though, all brains do this, not only German brains.

A mental timeline

REMEMBER, IN ORDER to create a feeling for time, the brain does not orient itself according to the actual amount of time that something takes. Measuring actual times would be quite arduous on the one hand. On the other, it is often unnecessary for the brain. It's much more important for the brain to organize experiences according to sensory connections and, in this way, to adapt the experiences to a timeline. But how does the brain go about doing this?

We learned at the beginning of this chapter that we often muddle our time planning for future actions since we can no longer correctly estimate how long these actions took us in the past. Routines, in particular, shrink in duration in our memories, always seeming to take a shorter amount of time than they actually did. The reason for this is that our brains do not measure time by minutes or seconds but rather by experiences. The experiences are organized into a chronological order on a kind of "experience timeline." This is not a temporal but rather a "living" axis. The more experiences there are on it, and the more intense the experiences are, the more time we allocate to each experience. First comes the experience, then comes the memory of it, and only at the end do we get the temporal perception.

One brain region that assumes this task is the insular cortex, located just above the temple, which is overgrown during the brain's development by the cavities and indentations of the cerebral cortex. One section of this insular cortex is apparently something of a gathering place for emotional moments and their temporal effects.[11] It creates a whole experience out of all of the incoming emotional conditions so that we are able to view ourselves as emotionally unified beings. This allows the brain to construct an emotional timeline for ordering experiences chronologically. Anything that has been experienced with particular intensity is covered extensively by the insular cortex, taking up a lot of space on our personal timeline (and thereby being perceived as having taken more time in retrospect). Boring experiences are given less room by our insular cortex and are allotted little to no space on our timeline. In extreme cases, we can't even recall certain boring episodes. It seems to us that they didn't even happen. All of this goes to show: time is not measured by our brains so much as it is artificially manufactured.

Why time seems to fly as we get older

BECAUSE THE BRAIN does not measure time, we should be particularly careful about relying on our perception of time. Our feeling for time is almost always wrong since it is oriented around our subjective experiences. This is also a reason why our feeling for time changes depending on our age.

When we reminisce about our childhood, certain phases seem to be much longer than they were in reality. The older we are, the faster time seems to pass. Weeks, months, and years fly by as if in a jet, and we can hardly keep up with our work and deadlines. There are two possible reasons for this. Either we truly have too much to do in the amount of time available, or our perception of time is off. Fortunately, the latter is more often the case. This opens up the possibility for us to reframe our subjective time perception so that we can "win back" more time.

The fact that time seems to move more quickly as we age does not mean that our sense of time has gotten rusty. Both older and younger people are equally good at estimating an interval that lasts seconds or minutes.[12] What changes is our subjective feeling. If both older and younger test subjects are asked how long it takes to cross a street, the younger subjects are much closer to the actual amount of time required. Older test subjects underestimate the time.[13] This is not only because we start to feel less physically fit and spry in our old age but also because our brain distills an entire compilation of past experiences down to a generalization. We have simply crossed so many streets in our lifetime that we can no longer recall every detail and thereby forget how long a crossing actually takes.

Because time is only organized in retrospect, our perception of time shifts during our lifetime. When we were younger our experiences were almost always new, or we were having them for the first time. In other words, we experienced far more memory-worthy things, and the density of experiences was much higher. But because our brain sorts and organizes these experiences only in retrospect, the time they took seems to expand in our memories. Later, when these same experiences occur more and more often, they are no longer new and are therefore omitted in our memories, which seems to thin out the density of our experiences. To put it another way, the saying "It seems like only yesterday" is nothing more than a disclosure of monotony in our own lives. But for someone who regularly tries out new experiences, the past will seem longer. We are hopefully able to recall our first kiss, which felt like it lasted an infinite amount of time in our memory. If you are someone who kisses frequently, you know that your memories of kisses seem to grow shorter and shorter. Perhaps your kisses really have measurably gone from minutes to seconds (ask your partner), but even if every kiss lasted an equal length of time, you would still emotionally recall your first kiss as the longest.

As one ages, time does not pass more rapidly and, in the present moment, it does not seem to be racing by us. But when we look back on it, we perceive the illusion that our experiences are more and more compressed because most of them are no longer new experiences. Our few but intense memories seem to follow each other very quickly because they are less frequent, and this makes them feel stacked up in our memories and therefore rushed. You might experience a similar effect when you regularly go to see your friends and then return via the same route. The way home seems to be shorter than it was on the way there.[14] It's the same when watching a movie. When you watch a long film the second time, you are surprised at how quickly it seems to end. If, on the other hand, you live a varied lifestyle, you will have a lot of wonderful events to think back on, which will stretch out the perceived feeling of time passing and make it feel not only fuller, but longer.

The advantage of temporal distortions

CLEARLY, THE BRAIN is quite bad at objectively grasping the timing of events. On the contrary, it actively defies temporal precision, sacrificing it for dynamic memories. But why? Especially when this only leads to more mistakes?

It turns out that one thing that is much more important than mathematical timeline perfection is to be able to store meaningful experiences from the past—in the present. Valuable lessons and intense experiences should not simply be plunked down on a mechanical timeline since doing so would make them less available for recall, as well as susceptible to being repressed by more recent, though less meaningful, experiences. Our sense of time as a by-product of our memory is necessarily subjective. It is only in this way—by giving space to intense and new experiences—that we are able to keep them vivid in our memories.

For this reason, the brain chooses to compress or expand periods of time as needed. The decisive criterion that goes into the decision is not the absolute time but rather the density of the experience. The more critical an experience is, the longer it seems to last in retrospect. Routines, by contrast, are recalled as accelerated experiences—"same old, same old," nothing new or exciting—in which case the brain compresses the experience until only a fraction of the actual time remains in memory. This allows important experiences to be remembered more intensely and less important experiences to be curtailed. The more boring one's life is, the faster it seems to go. A practical trick of the brain, I think.

Congratulations, you've gained some time

IF THESE NEUROSCIENTIFIC assumptions are in fact true, then it isn't all that difficult to gain a bit of extra time for yourself. Firstly, you need to rid yourself of the notion that you are going to recall with absolute precision the amount of time you have invested into an activity. If you spend double the amount of time on something, you will not remember it that way (and if it is particularly boring, you may even recall it as having lasted only half the time, or not at all). If you tend to have the general feeling that life is speeding up all the time and that you are unable to keep up, you might want to ask yourself why your brain presents you with this impression. The reason is because your experiences are too mundane for your brain. There's not enough new stuff going on! If you have managed to compress your daily routine into a highly efficient schedule, your brain is correspondingly going to shrink your sense of time for these activities into a highly precise, compressed little block.

Time cannot be gained in the present moment but only later when the brain reconstructs its sense of time for your experiences, which by the way, will feel longer the newer and fresher they are. If you intentionally interrupt your daily routine with wacky activities, I guarantee you will be surprised at how long these experiences will last in your memory. You could watch *CSI* for an hour in the evening and spend your time waiting to see whether the show's criminal investigators are able to crack their case (they will, though I won't spoil it too much for you), or you could invite five friends to play the card game Mau. If it is a fun evening, time will seem to fly by during the game night, but when you later look back on your evening, it will feel as though it lasted much longer than your average Sunday evening couch-potato crime-show routine.

Those activities that we find particularly fun always seem to fly by. And the opposite is also true: when something seems to pass by in no time at all, we remember it as something pleasant even if it was actually monotonous. This finding was revealed in a study in which time was manipulated for participants without their knowledge. Two groups were given the monotonous task of crossing out words with double letters (for example, the word "letter") for ten minutes. But one group was later told that they had only undertaken the task for five minutes, while the second group was informed that they had been crossing out words for twenty minutes. Interestingly, the second group, who felt that time had passed more quickly than it actually did, rated the same boring task as more positive and pleasant than the first group.[15] Does this mean that the opposite might also be true, that not only do nice experiences seem to pass quickly, but that activities which pass quickly feel nicer to us? Perhaps experiences only become valuable when time seems shorter...

When it comes to stressful work deadlines, for which there never seems to be enough time, it's unlikely that this will ever really be the case. But what about other, private, day-to-day experiences? People often wish that they had "more time" for the good things in life. But strictly speaking, what they want is something else: to be able to experience the moment more fully and abundantly. It's a paradox then that the most wonderful moments in life are those which seem to be over too soon. Our most enjoyable experiences hurtle past as though we are horses on a racetrack. If we had loads of time for everything, wouldn't that dim the power of wonderfully intense moments? A lot of people claim to desire immortality. But I imagine that were humans able to achieve this dream, they wouldn't be all that happy. Because once you have eternal time on your hands, why would you treasure a single moment?

Instead, ask yourself: Would you really prefer for time to run more slowly or would you rather have pleasurable and exhilarating experiences? If you choose the latter, you can be certain that you will always have too little time for memorable moments, as time will seem to fly by. But intense, racehorse moments are those which will later be given the most ample space in your insular cortex. Fortunately, your brain deceives itself by ensuring that anything that is diverse, exciting, or simply makes you happy is remembered as something long lasting. So, you don't really need more time in the present moment. You only need more intensity and variation and less automation and routine.

Time is gained by turning the brain's weakness into a strength. Instead of letting your brain reduce the most mundane and routine activities in your memory, do something entertaining and fresh that your brain will prolong in retrospect. The more routine your life is, the more fleeting it will seem. It's no wonder we sometimes arrive too late at meetings because we underestimate time. In addition to our time-estimation problems, there is another aspect that might be affecting this tendency for being late. Namely, what is worse for our brain than arriving too late? Arriving too early and having to wait. Our brain avoids waiting like the plague. Why? Stay tuned, the next chapter will tell you why.

6

BOREDOM

*Why We Have Trouble Switching Ourselves
Off—and How Daydreams Evoke the Muse*

S o, ENOUGH READING. It's time for a bit of relaxation. And
what could be better than, at last, to kick back and unwind?
No problem, you've earned a moment of peace. Sit down
somewhere comfortably, breathe deeply in and out, and fill
your brain with fresh oxygen. Shut your eyes and think about:
nothing.

Don't keep reading. Keep your eyes shut! Keep thinking
about nothing!

I'm not entirely sure how I should expect you to keep reading
if you do, in fact, have your eyes shut, but I do trust that your
curiosity will eventually get the better of you and that you'll want
to find out what happens next. Because thinking about "nothing"

is actually not easy, unless of course you are a meditation expert. Various ideas constantly rise up and form the strangest lines of thought: *"Now I'm sitting here. Should I open or close my eyes? If I keep my eyes closed, how can I read? This author is not very reasonable. Although with my eyes shut I am better at listening. Oh, there's a car driving by just now. I have to get gas for my car. Gas stations smell funny. I need to take a shower."*

The brain can do so many outstanding things. But it can't do nothing. We either think or we do not, and if the latter is the case, we are dead. Which is not a pleasant thought, not even for someone who is no longer alive to think. In principle, there is no single condition in which our brain is clear of all thought. Even when we are sleeping, it's not lounging around on its lazy meninges but is permanently active.

Shutting off is not really a strength of the brain, and this can burden us in two ways. Firstly, real periods of boredom are very unpleasant for us. Anyone who claims that they want to do "absolutely nothing" during their vacation has not really thought through how painful and awful doing nothing actually is. There is no task crueler for the brain than total boredom. And as we will soon see, the brain starts to make the wildest efforts in order to avoid such a "bore-out."

Secondly, our constant waves of thought often put us into a bad mood during downtime. Not being able to let go means that our unsolved problems and conflicts bubble up in our minds whenever we have a moment of peace and would actually prefer to relax. But our brain is unable to relax and so it not only plays back ideas, it also replays our problems over and over again.

For these reasons, boredom and daydreaming have a bad reputation in our society. Rightly so, since anyone who has so little to do that they can afford to mentally check out is not deemed all that productive. But. Wait just a minute. In reality, the mind's wandering is one of the greatest strengths of the brain. It is this propensity to daydream that makes it possible for the brain to take enormous leaps and to break out of fixed thought patterns.

In this chapter we are going to look at both the bright and dark sides of doing nothing. As irritating and strenuous as it might be, doing nothing can also help us to free up new ideas. The art lies in changing your perspective, thereby transforming negative periods of boredom into positive leisure time.

Our basic mental setting

THAT THE STRESS of the world cannot be released merely by pushing a button but rather continues to gnaw away at our thoughts is due to our basic attitudes. There is a region in the brain specifically responsible for triggering a "basic mental setting" whenever you aren't doing anything.[1] We are thus always thinking, basically. It's not as if our stream of thought could suddenly just stop so that a randomly passing neurologist might declare us brain dead. No, the brain can't take that risk and so it has created a mental empty mode, a cerebral "idle" state, if you will. This is why this particular neural network is called the "default mode network."

This curious region of the brain was discovered when researchers studied the mental activities of participants through a brain scanner. They stuffed volunteers into a magnetic resonance imaging (MRI) system and were able to determine, through a complex procedure, where the blood was flowing through the brain, signaling an increase in energy metabolism that indicates a concrete thought process. Don't get me wrong:

we can't read the content of specific thoughts, but we can see how the brain distributes the blood flow for its thoughts. It's not unlike Germany's Oktoberfest—wherever the most beer is flowing, the mood is usually the most buoyant. Theoretically, you could also predict that the tent in which the most pretzels, bratwurst, and beer are being consumed (equivalent to the energy metabolism) also contains the loudest guests. You see that this kind of "measurement" comes with a bit of a delay, but this is still similar to the way we measure the "activity" of neurons in the brain—indirectly and with a two-second delay through the blood flow.

As a scientist, one faces a problem at this point because if you want to measure thought activity in the brain, you are going to need a comparison value. Consider that you want to measure what is going on in someone's brain when they look at a photo of Dirk Nowitzki of the Dallas Mavericks. Naturally, you would first show your test participant a photo of the German "Wunderkind" basketball player while they are inside the MRI scanner, and then you would register which region of the brain experiences higher blood flow (in this case it will be concentrated in the neck area where image processing is located). But in order to do this, you first have to take a look at the level of brain activity that exists during a resting state so that you can make a comparison.

Experience has shown that this mental "white noise," the state of "nonthinking," is more common than people might like to admit or even realize. In order to obtain a basic comparative value, test subjects are shown an image with a small cross (so that they know where to look in the incredibly loud magnetic resonance tube). But while gazing at the cross, the brain doesn't think about nothing. Instead it wanders around with its thoughts: *I'm going to see a picture of Dirk Nowitzki soon. Basketball isn't really my thing. But he has always played for Dallas in the NBA. They seemed to set a new bar for losing on the court last season. Maybe that's why they're called the Mavericks?"*

In short, even if we aren't thinking about anything in particular, there's quite a lot going on inside our brain. It was especially interesting to learn that these basic white noise brain regions look the same for all humans. They are an amalgamation of all the areas of the brain that make up the default mode network and which fire up whenever we are "doing nothing."

This combined network is a collection of the crème de la crème for mental rumination: lateral brain regions that serve to integrate language and memory; central regions that internally focus our attention;[2] and one region that acts in a sense as a switch between our internal and external perception, called the precuneus, which is located just behind the crown of our head.[3] These brain regions are also involved in reflections on our own thoughts and are able to explain why we cannot consciously daydream—namely, our consciousness is taken over by the default mode network, and we usually only notice toward the end of daydreaming that our minds had been wandering.[4]

Better to be shocked than to wait

OUR BASIC MENTAL mode is thus: even if we aren't required to think any concrete thoughts, we still have to think. What we think about doesn't really matter. Or to put it another way, we are simply incapable of flipping the off switch on our thoughts. However, when we feel forced into this situation, things start to get really annoying.

If you don't have anything to do, it doesn't take very long for you to get bored. For a long time, the mental state of boredom posed a mystery for scientists. Boredom is not an easy topic to study. Experiments were undertaken by showing participants a boring video (for example, a dripping water faucet or a soccer game with no goals scored) and then studying the participants' mental capacities by administering them concentration and attention tests. But during these studies, researchers became aware that the administered tests were much more boring for the participants than the already quite mundane videos. Because what really bores us to tears, more than anything else, is having to do a monotonous task. Anyone who has ever needed to fill out forms or spreadsheets knows this. If you really want to bore someone, you should give them a task that is as repetitive as possible, such as proofreading a text, sorting thumbtacks, or screwing together nuts and bolts. Or you could isolate them completely from any type of activity and simply put them alone into a room where they are forced to wait.

It is exactly during a moment of forced meaninglessness or a nonsense task that the default mode network in the brain switches on. In other words, our inner thoughts resurface and start to wander. And this is apparently not a very comfortable feeling for people. If you put participants in a room for fifteen minutes and offer them the option of either doing nothing or administering painful electric shocks to themselves, two thirds of the male participants, and even one fourth of the female participants, choose to be shocked.[5] How weird is that? After all, we are carrying around with us a brain that is not just any brain but the most advanced brain that exists in nature! In the time span of fifteen minutes, we could compose a beautiful melody or embark on a glorious journey of the mind. But instead we reach for the nearest stun gun and give ourselves a couple of electric jolts. And why do so many more men than women choose the electroshock option? Maybe the primary thoughts of men are so horrific that getting shocked is the better choice? No one quite knows for certain.

It's much easier scientifically to study why people who are watching a boring movie prefer to eat junk food. Test subjects who were required to watch a repetitive film clip (about one and a half minutes in length) for one hour doubled their consumption of chocolate candies (incidentally, the number of self-administered shocks also doubled in parallel with this experiment—to an average of one shock every three minutes).[6] The reason for these snack attacks is not that participants were trying to generate a positive feeling for themselves by grabbing for snacks and sweets. Rather, their decision has much more to do with their desire to avoid the uncomfortable feeling of boredom. In this situation, candies or chips do not represent a tasty reward but rather the only possible way of getting through the soulless punishment of a bleak task. One could also say that someone who is constantly chomping away during a film isn't really absorbing the storyline. Since learning this, I am always careful now to check out the snack counter before I go to the movies. The more goodies the movie theatre sells, the worse I think their films must be.

In summary, several studies show that boredom and a bad mood go hand in hand. When you are forced into an undesirable waiting situation, you are suddenly confronted by your own ruminations and problems that start bubbling up and waltzing back and forth in your mind. We are unable to turn our brains off, and this inability makes us unhappy since most of the thoughts that we have when daydreaming have a tendency to dampen our mood. This finding came out in 2010 by means of an iPhone app that regularly asked over 2,200 participants about their current mood, what they were doing at that moment, and what possible daydreams they might be having.[7] The results revealed that when we do a task, nearly half of the time is taken up with mental distraction, instead of actually thinking about what we are doing. In addition, daydreams and mental rabbit holes more often than not lead to a downswing in our moods. One exception was when the test participants had just been occupied with having fun with a partner in bed—but who in that situation wants to be distracted by their thoughts or by an iPhone app? Although... it depends on the situation. Almost half of German women reported that they would choose to abstain from sex rather than abstain from using their smartphones for an entire month.[8] This is perhaps no great wonder, however, if one considers that the user-friendliness, accessibility, and organizational skills of smartphones are unrivaled, but that these same features differ according to each individual model when it comes to men.

Bored? It's your own fault!

SO, OUR BRAINS switch to "daydream mode" not only in boring situations but also during almost 50 percent of all practical activities and are almost always in a worse mood afterward. Dwelling on our own thoughts is thus not necessarily a good thing. But the primary weakness of our brain is its inability to shut off our thoughts. This inability is responsible for our eventual ruminations and concentration errors. Boredom reinforces these tendencies, which is why it is seen as uncool and a stigma in our society.

Our workplaces add to this by offering the perfect recipe for a well-rounded state of boredom, including workflows meant to be carried out efficiently, with as little variation as possible, in order to increase productivity, along with monotonous processes that suppress any form of deviation. Only those things that can be shaped into a controllable and measurable schematic are considered truly profitable. However, standardized processes that minimize errors and distraction simultaneously create the perfect conditions for boredom. Because humans aren't made for highly efficient workflows but require variation in order to maintain concentration. This is a problem (generally difficult to measure) in our modern form of intellectual work. Because as long as our performance continues to be measured by work hours, only those employees who are the last ones to leave at night will appear to perform better than their colleagues— regardless of whether they spend the last hour before going home scrolling around on the Internet.

Anyone who is bored gets quite a bad rap nowadays. Contrast this with someone who suffers a well-earned case of burnout. These days, it seems that anyone who is so conscientious and passionate about their work that they collapse from being overworked can be forgiven for a period of temporary weakness—as long as they return to the job even stronger and more efficient than ever before. The modern breakdown seems to count as the mental catharsis for furthering one's career. A burnout, oh yeah, that's a badge of approval for the devoted worker. A mental scar to be displayed like a trophy as one scuffles up the ladder. The opposite of this, the bore-out, spurred on by the anguish of eternally monotonous tasks and frequent mind wandering, enjoys much less respect in our culture.

But there's more to boredom than this negative aspect. The ability of our thoughts to wander is in fact a human strength, which enables us to solve problems in a new way. This goes in the face of our popular ideas about productivity. These days, anyone who hopes to advance in their career at some point attends a seminar with a title like: "How to think more efficiently in your workday!" or "How to push your concentration to peak performance." I have yet to see a workshop, book, or seminar entitled: "Boredom in 10 Easy Steps!" or "A guide to effective boredom!" But why not? Because within boredom, which can be very painful for your brain to deal with, there is also creative potential and the possibility of combining so many ideas in new ways. In fact, it contains exactly the thing that our modern world is always demanding. I can see it now, the professional boredom course: "Come on in and get bored as you never have been bored before." But no. Instead, we consider someone who gets bored to be a loser, a failure, a person who doesn't have anything better to do. We think that anyone who is just hanging around must not be producing any kind of measurable service, or is not living up to their talents, but seems to be wasting them. Or is it possible in some cases that it could be the other way around? Perhaps a person has so much to do that the really important ideas only occur to them whenever they are bored?

Idleness is the root of all ideas

ALL RIGHT, I'LL admit that "getting bored" is perhaps the wrong term to use, since boredom, as we have just read, has a bad reputation and for good reason. It is most often something that happens against our will. We usually don't choose to be bored for half an hour but are forced into it while waiting or for some other reason that keeps us from doing what we actually want to do. Boredom is therefore something like the younger and meaner sister of that grande dame of idea incubation: the Muse. Worshipped in ancient times for their inventive powers, the Muses were deified by the Greeks. This seems rather appropriate, in fact, since we know in neuroscience that a somewhat relaxed, Mediterranean attitude to work is a valuable component to thinking well. Musing and idleness go hand in hand. Consider that the origins of the word "muse" refer to wasting time, or literally, to standing around with your nose up in the air. Idleness may be the root of all evil, but it is also the beginning of all creativity. Someone who works nonstop and as productively as a highly efficient industry robot is, at the end of the day, only as mentally capable as the same robot. Just ask Siri, the voice of the machine world. When I ask her the question: "How important is it to take a break?" she answers: "I don't know any app for that. But would you like me to search in the App Store for 'take a break'?" No, Siri, you've got it all wrong! Frittering away the last remnants of one's free time on an app is *not* the solution. Doing so would force us to sacrifice one of our real strengths— namely, that we are able to let our thoughts meander whenever we take breaks. No machine is able to understand the bright side of inefficiency.

The fact that our brain is unable to switch off but keeps bouncing around among its own thoughts has a positive side to it. Participants in research studies are often bored to tears by the dullest activities (and, I should note, frequent boredom seems to have a negative effect on life expectancy),[9] and yet people claim to find relaxation, contemplation, and even joy in seemingly monotonous tasks such as knitting, coloring, or meditation. The reason for this is clear. Self-selected moments of withdrawal have a qualitatively higher effect than moments of unwanted, imposed boredom. And while the daydreaming brooder is secretly ashamed of having whittled away their time on meaningless mental loops, the knitter feels refreshed and liberated. Even if no one else wants to try on the scratchy sweater.

Whether we find ourselves in a condition of boredom or in a moment of musing idleness, the differences it produces in our brain are small but significant. Boredom is when one has the possibility to do something, but the situation in which one finds oneself is so tedious that one can't make oneself do it. Some of the regions of the aforementioned default mode network that are usually responsible for spurring us to take action are less active when we are bored.[10] In theory, we could do something to get over our boredom, but we are too bored to bother. Periods of idleness, on the other hand, are when rather than being externally confronted by monotony, we can decide for ourselves to have productive downtime. As a result, we are able to actively interact with our surroundings, thereby giving free reign to our thoughts.

Making the best of our inability to switch off

OUR BRAIN IS unable to switch off, meandering here and there with its thoughts as a result of our default mode, and there's not much we can do about it. Yet such a time-consuming and fixed feature in the brain cannot merely be a coincidence. There must be some advantages that could explain this type of behavior. Why else would the brain simply swallow all of the disadvantages lying down—the inability to switch off, all the ruminating, the daydreams and everything that they lead to?

If we were unable to concentrate on anything other than the present moment, we would be nothing more than biological automatons, capable only of knee-jerk reactions to the demands of the moment. But we are not like that. We are, in fact, capable of detaching ourselves from the present moment and taking on a variety of perspectives. We can travel through time in our minds in order to reach better decisions. We are able to empathize with others and thereby better understand our fellow humans. We can remove our mental blinders and land on fresh ideas. Our long-neglected default mode network appears to be something of a mental secret weapon, helping to release us from the reflexive question-answer pattern of our surroundings and offering us new perspectives.

Productive daydreaming

WHETHER OR NOT our default mode network is able to productively process our thoughts depends largely on the nature of the task. The more concrete the task or activity, the easier it is for wayward thoughts to distract us from what we are supposed to be doing. For example, if we are concentrating on reading a book while simultaneously wondering about a strange look that a colleague gave us earlier in the day, it is going to be hard to remember the text. Perhaps something like that has already happened to you while you were reading this chapter. You may have been reading along and then suddenly realized that your thoughts were somewhere else entirely. But whereas this might make the average author feel sad or distressed, I personally feel happy to hear it. Because it shows that your own thought-journey function is working beautifully. I'm not very keen on the kind of reader who focuses their attention on every single word. I prefer those whose thoughts stray every now and then. They are usually the most creative types.

But let's bring our thoughts back to a focused reading of this text once again because there is some important stuff here. You might be surprised to know that while we sometimes allow our mental wandering to distract us from a concrete task, this kind of distraction is actually desirable in other sorts of tasks. When it comes to so-called open tasks, which require little stress or hard work, we're less likely to be distracted. To put it another way: if you want to recite the lineup of the Dallas Mavericks in the last season, you'd best focus and not get distracted. If, however, you are asked how to change up the squad's lineup so the team will make it to the playoffs next time, your best bet is to open up your mind to a bit of distraction and allow strange new ideas to start flowing.

It is important to equally balance both concentration and distraction and to fit them to the task. People who practice this deliberately are not only more creative, but they also tend to make fewer impulsive decisions. This makes sense since someone who can play out in their mind the possible consequences of their decisions will be much less likely to go along with a rash course of action. These productive "dreamers" are able to hold out for a later reward and to thus make better decisions in the present moment.[11] If you want to keep the bigger picture in mind in the midst of a stressful and emotionally charged situation, you must pause for a moment, move in your thoughts, and take a mental step back. The default mode network is what enables us to do this.

When we think creatively, the dual nature of distraction becomes clear. New ideas are only generated when we have loosened ourselves from a concrete task and are able to subconsciously work out possible solutions. Our "daydream network" assists us in this undertaking. When participants are asked during a classic creativity test to come up with as many possible uses for a brick (or some other everyday object) in one to two minutes, results show that new ideas are better generated when there is a break in the creativity test, and only when the participants daydream during the break. This can be achieved in a laboratory setting by asking the participants to complete a monotonous task for a short period (such as crossing out letters in a text for a couple of minutes). But if the participants are forced to wait in boredom for the duration of the break without any kind of task, they display as little creativity as they would if they were required to perform an activity requiring concentration during the break (such as memorizing a poem).[12]

Note: Whether our brain switches into an awful boredom mode or into a creative period of leisurely daydreaming depends on the environment. If the goal is to solve a creative task (or to reach an abstract and sweeping decision), intermediary, monotonous tasks will help to generate the state of muse-like idleness, rather than boredom. In the end, this is precisely the phenomenon of idleness, or leisure: finding the balance between relaxed inactivity and mental wakefulness.

If you have to be creative for too long, the benefit of daydreaming passes. Imagine that you are instructed to come up with as many new possible uses for a brick not for only one minute, but rather over twenty minutes. Regardless of how many daydreams you've enjoyed previously or how many muses you've been touched by in your period of leisurely idleness, you won't be able to keep up a flow of fresh ideas for twenty minutes. Our thoughts will start to stray after only three or four minutes at most.[13] Our daydream network pushes in because it wants to give us a mental break, and so our thoughts start wandering again, leading us away from our original goal. This serves to show once again that our default mode network is only able to display its strengths under the right conditions. Make sure you pay attention whenever you want to activate your brain's daydreaming power. The point isn't to take a break in order to stimulate your thoughts to wander. It is only when you have a long-term decision to make or need to solve a problem in a new way that the power of daydreams can be usefully harnessed. If this is the case, first focus your attention intensively on the problem at hand for a short period of time before allowing your default mode network to switch on your daydream mode, thereafter returning with renewed concentration to your task.

The art of interplay between tension and relaxation

THE FACT THAT thoughts are constantly swirling through our brains, affecting our mood or causing us to brood, is the price we pay to be able to understand ourselves and to consider unfamiliar perspectives. Just imagine the alternative: without the freedom to digress mentally, we would be nothing more than intelligent robots automatically processing sensory stimuli. Highly efficient, if extremely boring, and always the same. The brain is not made for this kind of work. We see this because the more monotonous our task, the more our brains wander. Mental daydreaming like this might stimulate creativity in the beginning but, at some point, we eventually get bored of this too and require fresh input. This keeps us curious and makes our lives diverse, thanks to our default mode network.

When we freely choose to undertake a task requiring little mental exertion, our thoughts are particularly well suited to go off on a productive exploration. It's no wonder that many people get their best new ideas when taking a shower, cleaning the apartment, or listening to music. Our society should take care not to undervalue measured downtime, a well-fostered period of idleness. But nowadays everyone carries around a smartphone, making it difficult, if not nearly impossible, to give our thoughts the mental space that they require. Our lives are not made up of a linear sequence of problems that need to be solved but rather of the interplay between tension and relaxation. And it is precisely this period of relaxation—of leisure and idleness—that is a valuable tool for generating the really important thoughts.

So, what should you do if you can't get your brain to switch off and are constantly plagued by unsolved problems running amok through your head? The answer is simple: deactivate your default mode network by engaging in a mentally demanding activity. I like to ride my road bike, but this is the wrong activity when there's something I can't stop thinking about. Because the more monotonous the activity, such as biking, the more uniformly your thoughts will circle back and forth around the same topic. A better alternative is to meet with other people in diverse surroundings because the more mentally challenging the activity, the fewer resources your brain has available for mental wandering. Allow external factors to distract you rather than your internal thoughts. You could also choose to meditate. This has been shown to decrease the activity of the default mode network.[14] Of course, this will result in less creativity, but when it comes to the brain, there is always a mental price to pay.

7

DISTRACTION

*Why We Are So Flighty and Which Distractions
Can Lead to More Creativity*

NOWADAYS WE HAVE a big problem. Because...

Wait a sec, I just got an email. I have to answer it real quick.

...because we are often distracted from our work. In principle, this is nothing new. The condition of being distracted from our work might be as old as work itself. At the end of the day, we are communicative beings from the ground up. The more opportunities we have to communicate, the easier it is for us to get distracted. We are living in what is probably the most communicative age of humankind. In any case, only a handful of years ago it was not yet possible...

Oh, a message from my sister in Australia. Hang on just a minute.

...not yet possible to tap on a glass touch screen in order to keep in touch with people anywhere in the world. A blessing and a curse, that's what it is often called, because whatever is able to capture our attention also makes us less productive at the same time. Distraction is the natural enemy of concentration.

It's no wonder then that interruptions are the number one performance killer at work. According to one study by the U.S. job network CareerBuilder, around one fourth of employees spend more than one hour a day sending personal messages or phoning friends or family—in other words, not concentrating on their work.[1] Unbelievable! An entire hour of the workday spent using one's own private social media! American teenagers (ranging from the ages of thirteen to eighteen) would probably only muster a weary smile about these results because this particular population spends an average of nine solid hours per day using digital media. This number doesn't even factor in texting, sharing, and liking during class time.[2] Statistically speaking, this means that teenagers spend more than half of their waking hours on distracting computer games, surfing the net, or posting messages. Seen in this light, one might argue that school and homework are annoying interruptions for your average teenager's consumption of digital technology. The same study also found that only 10 percent of teenagers reported social media sites such as Facebook or Snapchat as their "favorite media." So far, the Silicon Valley giants don't yet have control of today's youth, over half of whose primary preference remains to loaf around in front of a television.

Whoever read the previous chapter with concentrated focus...

Another buzz from my cell phone. I have to get that, excuse me.

...with concentrated focus and attention (which is no easy task, I know), will be able to recall why this is so: boredom and monotony are like poison to our brain function. Our brain always wants to experience something new. But there is another reason, which we should add here. Our brain's filter mechanism, which is normally responsible for suppressing distracting interruptions, is cleverly tricked by our workday routines and the use of social media. This leads to a loss of concentration as we give in to the distraction of a buzzing phone. On the other hand, it also shows that our curiosity, our search for change, is a powerful driving force for our brains. It is only by way of distraction that we are able to think outside of the box.

Like many of the brain's features, its penchant for distraction is a double-edged sword. It helps us to grasp new ideas while at the same time it blocks our productivity. So just what can be done to exploit the brain's innate weakness for distractedness and to protect its openness to new inspiration on the one hand, and on the other hand, to be able to concentrate when concentration is called for? In order to answer this, we have to understand how the brain weighs information and releases it into our conscious awareness. I promise not to get distracted any further throughout the next pages. But before we move on, I just need to check my email inbox.

A spam filter for the brain

SPEAKING OF EMAILS...Before you can productively begin to communicate via email, you usually have to do a little housekeeping—namely, clean out your inbox, which has been cluttered up with spam. According to recent analyses by cyber security firms such as Norton or Kaspersky, over half of all emails sent are useless or even harmful trash.[3] In fact, 50 percent of global Internet traffic is sent by automated programs (so-called "bots") that direct annoying queries to websites or build up spam email networks.[4] Even computers get annoyed at interruptions from their colleagues. Fortunately, there are filter mechanisms that act as a shield from such exasperating emails. In my email program, I simply click on the setting "Filter spam" and all of the irksome Viagra offers and dubious bank account inquiries quietly disappear.

If you look closely enough, you will notice that most email programs offer multiple ways of filtering out spam. You can reject the spam immediately and never even allow it into your mailbox. Or you can collect it first in a spam folder and then later browse through the folder to be sure an important message has not accidentally fallen prey to your spam filter. Maybe you will one day want to look more seriously into those Viagra offers, who knows?

Similar brain filters were already studied in the 1990s,[5] long before spam even existed. Similar to the manner in which a spam filter can either delete spam emails as soon as they arrive or store them in a folder for later perusal, the brain also filters out sensory impressions either immediately or retains them until a later time when they will be discarded or used. In principle, we thus have two filters in our head: an early filter which promptly blocks sensory impressions and a late filter, which decides whether we might later become aware of a potentially important bit of sensory stimulus. However, the brain is not like a static email filtering system for which the mode must be determined up front. The life of the brain is more multifaceted than that of a computer, and its filter must adjust depending on our level of mental exertion. In other words: whether something distracts us or not does not depend so much on the intruding stimulus as it does on how hard our brain is working. Our brain has a highly capable spam filter or, er, I mean stimulus filter.

The brain's portal

THE BRAIN'S EARLY filter works to immediately block out intrusive interference. We are not even aware of some stimuli because they are actively shut out. This filter is located underneath both of the brain hemispheres in the thalamus portion of the diencephalon. The diencephalon regulates a large part of our subconscious bodily functions such as our heartbeat or body temperature. But the thalamus is further responsible for sorting incoming information. True to its Greek name (*thalamos* means "chamber"), it acts as the portal to the cerebrum. This is what decides which information is allowed to come inside and which is barred entry. Most stimuli are required to pass through this portal and are then corralled into the right direction in the cerebrum. Important bits of information are taken to our consciousness and unimportant bits are sent to the brain regions that work subconsciously and do not bother anyone. Almost.

Everything that is perceived by the brain (with the exception of smell, which is processed in the olfactory part of the brain) must pass through this information filter. The thalamus gets bored very quickly. Or, to be more precise, it is the king of boredom. If a sensory impression does not change at least every two seconds, it loses any conscious attention and is sent into the subconscious. Like a sharp, high-speed spam filter, the thalamus weeds out incoming stimuli if they are always the same. In other words, the importance of a piece of information is not determined by its contents but rather by its variation. This explains why we are constantly reaching for our smartphones whenever a new message pops up. The contents of the message are not nearly as interesting for our brain as the fact that something just changed. Change makes information exciting.

Filter overflow

THE THALAMUS IS thus something like a line of defense against intrusive distractions. But the brain has another filter in its arsenal as well: our working memory. Similar to how we can decide whether to immediately delete incoming emails or to hold them in a spam folder, our brains can also hold certain pieces of information on call, just in case we might need them later. This takes place in our working memory, which is able to store information for a few seconds. This also implies that the smaller our working memory is, the more quickly this particular late filter gets overwhelmed, thereby allowing us to get distracted. In this case, it is as though our spam folder got so full that it overflowed into our primary email inbox.

The less we are able to retain, the more easily distracted we become.[6] In a study that looked at participants' general memory retention abilities (i.e., how many objects or words in a list they were able to retain) by testing how easily they were distracted during a concentration test (for example, how often they took note of annoying pop-ups during a search puzzle), participants who were unable to remember many items on the list performed poorly. Conversely, however, even the best memory artists get distracted when their mental storage capacity is overwhelmed because they are no longer able to prioritize between important and unimportant pieces of information.

We notice this sometimes when we are reading difficult texts.[7] Hopefully, that isn't the case for you right now. You are reading along and have to take note of a lot of different things. Sometimes it gets to be too much. Your memory gets overwhelmed, and your mind starts to wander. Suddenly you realize you are somewhere else with your thoughts because you noticed a passing car or a conversation in the next room. Whenever your working memory gets overwhelmed, it becomes difficult for a new stimulus to be classified as relevant or unimportant. In this case, the brain says: "better get distracted before you filter out something else important." Texts that are hard to read are therefore best broken into chunks so you can give your working brain the chance to sort through them. Paragraph by paragraph or chapter by chapter, you can briefly take note of the contents and the most important messages and keep your working memory in shape for the next round of information. Right now, for example, would be the perfect moment to try this out.

Don't listen to your heart!

OBVIOUSLY, WE HAVE developed very sophisticated filter systems to protect ourselves from useless sensory input. And although at first glance it may seem that we get too easily distracted in our lives, our built-in early filter in the thalamus is so powerful that it sometimes even blocks out things that are very important for us to notice.

The most permanent and also most boring stimulus that the brain filters out is . . . our own heartbeat. Although the heartbeat enjoys a positive reputation and passionate praise among art, poetry, and music factions, there is in fact nothing so irritating for our brain as the steady, monotonous tapping of our heart. Before our brain was able to consider anything consciously, our heart had long since begun to beat. Our neurons have never had a quiet moment to relax because our heart has always been at it, pounding away in the background. Fully unfettered by information and tedious. Terrible. We have to assume that this monotonous beat is not a pleasant feeling—at least not for the brain.

No wonder that our brain has developed an effective system for suppressing the sound of our heartbeat. Interestingly, this heartbeat filter even gets applied to the outside world whenever the rhythm of our heart reoccurs in our surroundings. In one study, participants found it difficult to identify flashing figures whenever they flashed in rhythm to their own heartbeats.[8] But when the figures flashed in a different frequency, participants were able to see them without any problem. Let that be a warning to all you fitness armband fanatics: forget about your own heartbeat. Don't listen to your heart! Your brain knows why it is consciously blocking out this monotonous stimulus. And as anyone who cuddles up to a loved one to listen to their heart knows: it's lovely—for a few seconds. Any more than that, and I guarantee you your brain will have had enough.

Gorillas in the lungs

IT IS NOT such a bad thing that we don't tend to notice banal things such as our own heartbeat or a ring that we wear or the glasses sitting on the ridge of our nose. The informational contents of these particular stimuli are limited. But the brain's filter mechanisms are far from perfect and sometimes cross the line in the same way that a spam filter will, on occasion, block a meaningful email. In psychology, this phenomenon is called "inattentional blindness" (also known as perceptual blindness).

By far the most famous study in this area is the gorilla experiment in which participants were asked to watch a video and count how many times basketball players passed the ball. While following the ball with their eyes, the participants did not notice a woman in a gorilla costume sauntering across the basketball scene.[9] Not inconspicuously or discreetly, but strolling slowly like a primate and even stopping to drum its fists on its chest.

This phenomenon not only happens to test subjects who are untrained in noticing details—even professional search specialists fall prey to this filtering trick of the brain. In another experiment with experienced radiologists, 75 percent of participants did not see the image of a gorilla that was hidden in an X-ray image of lung tissue, even though the gorilla was forty-eight times bigger than the tumor they were tasked with finding.[10] Let me restate that: a gorilla that was forty-eight times larger than a tumor went unnoticed by specialists looking at an X-ray. And this even though eye-tracking tests showed that the doctors looked directly at the gorilla. It is fortunate for all of us that having apes in our lung tissue is a relatively rare condition.

Unbelievable, one might say. But the gorilla basketball experiment was conducted in 1999. No one could possibly guess that we would be even crazier nowadays. If you want to see what I'm talking about, you need only type "texting while walking" into the YouTube search bar to see how well our built-in filter is able to block out our surroundings. The videos that pop up depict smartphone users staring at their displays while unexpectedly walking straight into swimming pools, falling over flowerpots, and plowing down unsuspecting senior citizens. Particularly affected are young people up to the age of thirty-five, a fifth of whom, according to a recent study, cross the street with their heads down.[11] What can be done? Putting streetlights in the ground to act as ground lights might be one solution (which is already being implemented in the town of Augsburg, Germany). Or building a separate sidewalk for smartphone users so that they don't hold up the rest of foot traffic (this is being tested in China). But what if you still keep looking at your screen and step in a pile of dog poop? The smartphone industry is already on it: an app allows your screen to become "transparent" so that you can see where you are walking while continuing to type your WhatsApp messages. Great, so what's next? A "window app" that allows you to hold your smartphone up to a window so you can see what is going on outside?

The workload blinder

I'VE INCLUDED THESE examples to show just how double-edged our attention filters can be. Whenever we concentrate intensely on one thing, we filter out other surrounding things. No, even more than that, we don't even actively notice them. Brain scans show that the portion of our brain responsible for processing background images remains almost completely inactive whenever we focus on something in the foreground.[12] This is because the filter function of our thalamus is actively constricted by the cerebrum when we are focused on finding details in pictures. The thalamus is connected to the brain through extensive neural connections, and if the brain decides to favor a particular sensory channel, it can actively suppress attention to the other senses. A bit like a boss who instructs their secretary that they don't want to be interrupted by anyone other than their family members.

The brain also maintains this kind of top-down control. You can actively keep your thalamus filter from receiving or suppressing sensory input. For example, are you by chance wearing a ring or a pair of glasses right now? If so, take note of how your ring is hugging your finger or the glasses are pressing down on your nose. You have just told your thalamus to allow these sensations through the filter, though it won't be long before it starts to block them out once again. The opposite is also true: the higher the mental load, the more the thalamus filters out, even unusual sensory stimuli such as the processing of tones.

Scientists can measure this perceptual blindness by asking participants to complete a complicated search activity (for example, looking for a letter in a group of similar letters, such as a V surrounded by a cluster of other letters: WIWVWWIW). The more complex the search activity, the less the participant is distracted by background noises or superimposed images. The result is that one not only becomes blind, but also deaf to external stimuli.[13] And there's nothing you can do about it. Remember that the next time you are talking to your partner while he is eagerly watching the latest football game. Because of psychological reasons, he can't even hear you. And if he doesn't answer, it's not because he's being rude; he simply can't help it. Watching football takes a toll—namely, all of the brain's resources.

By now it should be clear to you that our mental filters are in no way as static as an email spam filter. Our filters adapt to our lives. The fuller our working memory is, the more sensitive we become to distractions. Even if the demands of a task decline, we will still be more susceptible to distraction. Conversely, this also means that there's really no point in spending your efforts on not getting distracted—your task also has to be sufficiently demanding.[14] But if we are intensely focused enough, we can be sure not to be distracted as long as our surroundings remain constant. In other words, it doesn't make much of a difference to the brain if you're in a soundproof isolation room or in a bustling open office setting with a steady flow of background noise. Both locations are equally productive for the brain as long as it is involved in a sufficiently challenging task. For someone finishing up an exciting project, it doesn't matter whether or not one's colleagues are chatting back and forth. If, on the other hand, you are filling out the same mundane Excel spreadsheet over and over again, even the slightest squeaky hinge will drive you bonkers.

The tricks of distraction

OK, FINE, YOU might say. The thalamus is so clever that it only grants entry to information if it is suitable to our workload. If that were true, we would never have any problems with our concentration as long as a task was challenging enough. But, unfortunately, there are many distractions that manage to avoid our brain's filters. Some stimuli are so interesting that they are even able to sidestep a hardworking thalamus. They use a mental bypass to escape being in the least filtered.

You are already familiar with one such bypass criterion: change. Monotonous stimuli are briefly registered and then quickly blocked by the thalamus. For precisely this reason, the basic settings of various types of media are particularly wily. Every ping, buzz, or flash provides a new change stimulus to the brain. After all, we reason, we don't want to miss anything exciting. This releases a cascade of thoughts in our brains that pulls us away from our current focus. In the previous chapter you read about the power of daydreams. This form of mental wandering can be initiated by something as small as a buzzing phone. Scientific studies have shown that a mere vibration of a phone effects our concentration as adversely as answering it.[15] Because distraction isn't so much dependent on whether or not we turn to a different activity (such as reading or writing a new WhatsApp message). It's enough that our attention is simply torn away from our workflow. The subsequent mental wandering severely restricts our cognitive performance by diverting mental capacity from the conscious control regions of the brain and devoting them toward that which has distracted us.

Emotionally positive stimuli are similarly good at tricking our attention filters. Study participants taking a test requiring intense concentration—finding letters among a group of similar letters—were more easily distracted whenever the images that flashed alongside the group of letters were considered pleasing.[16] The study was able to show that pictures of naked women were twenty times more distracting than mutilated bodies, and happy faces were more distracting than angry ones. Faces, in general, proved to be the key for capturing our attention, as though they hold VIP access to our brain. When participants in another study were also asked to find a letter among a group of other letters, they were much more distracted by pictures of faces that were flashed alongside the letters than they were by pictures of musical instruments—especially if the faces were those of famous people.[17]

What distracts us the most? If your smartphone vibrates and you look down to see a picture of Donald Trump's comb-over flapping in the wind or German chancellor Angela Merkel flashing a rare smile, there's not much you can do. You're hooked.

Training your portal

WHAT DOES ALL of this imply, in practical terms, about the ability of our distraction filters to function? Do we have to try minimizing background noise, do concentration exercises, or get rid of our smartphones?

The answer is: it depends. You've just seen that the most distracting stimuli originate in our brains and not from irritating external factors. You can turn your filter's efficiency up or down by the type of activity you are involved in. This becomes easy to see, for example, in research that studies the effect of noise volume in a classroom (or in an open office space) on performance. Noise pollution in workspaces is bad for health reasons, but it has also been said to have negative effects on workers' concentration, as well. A study from 2013 correlated the school performance of French students (between the ages of eight and nine) with noise levels in their classrooms and bedrooms. Scientists found that the louder the noise volume, the worse the students performed on average in French and mathematics. Anyone who studied in a noise level around ten decibels louder than the others (equivalent to double the acoustic pressure than normal) received on average 4 percent fewer points on tests in both subjects.[18] Note: the study confirmed a correlation, not a cause-and-effect relationship. Because, and this is the key point, volume alone does not create distraction. It must also fluctuate. It is this fluctuation that the brain finds difficult to withstand.

What would happen if you suddenly had to concentrate more, for example, because the text suddenly changed and became harder to read? One Swedish study from 2014 was able to clearly show that when this happens, audio noise suddenly became less distracting for participants. In the test, students were asked to read a text written in an unfamiliar and somewhat difficult-to-read font and then to read a second text in a familiar and easy-to-read font. During the reading of both texts, distracting background noise (a low background murmur) was played to interrupt the students' concentration. The distracting babbling noises worked, and the students retained less from the contents of the easy-to-read text. But while they read the text that was written in the less familiar font, the background noises showed no negative effects. The students were able to retain as much information from the difficult-to-read text as when they read the easy-to-read text in a quiet environment.[19] In conclusion: whether or not something distracts us depends mostly on our inner settings. To complain about one's surrounding environment is to bark up the wrong tree. Instead of changing your environment, try changing your mental task. There is no normal (less than eighty decibels) sound that is able to distract us as long as we are really mentally challenged.

Productive distraction

AN AVERAGE WORKDAY for many of us contains a lot of monotonous and similar tasks, which are therefore particularly susceptible to distraction. It's easy for me to be clever and write that we should simply make our work more exhilarating. Another, more realistic, alternative is to minimize potential distractions during mundane tasks by using the biggest weaknesses of our filter mechanisms—their weakness for change.

Someone who varies their tasks during their workday (for example, switching back and forth between writing and making phone calls) is already able to create a distraction *within* their tasks. The important thing is to switch up your work while continuing to do the things that need to be done. This helps the thalamus to learn how to prioritize and weigh information while continuing to remain on topic.

It is equally important not to overwhelm the filter mechanisms of your working memory. Working steadily without a break will eventually fill your memory capacity to its limit, causing you to become unconcentrated and distracted. Cognitive tests show that this usually takes place after thirty to forty-five minutes. But before the inevitable distraction occurs, you can jump the gun and "distract" yourself by taking a break. A clever break, that is. Instead of rushing to tackle a different problem, you should rather pause briefly, stand up and chat, thereby giving your brain the opportunity to mentally digest. A few minutes will suffice for such a break. Any longer and your brain will switch over into an ineffective distraction mode, making it harder for you to ramp back up to your working rhythm and to once again refocus on your work. Make your breaks short and snappy—but take them more often.

There is really no way to avoid the urge to distraction. Distraction is the logical consequence of our limited working memory. Our brains need a breather. But if you are able to use these breaks wisely, you might be surprised that it can increase your productivity, even on tasks that are dependent on efficiency.

In 2009, the Bank of America had a problem. The productivity levels of the bank's call center employees varied considerably. Some teams were able to tackle customer inquiries on the hotline rapidly, and seemingly without problem, while other teams lagged significantly behind. In order to improve the efficiency of the slow teams, a strict schedule was implemented and coffee breaks were regulated so that the team members had to take turns having breaks. This was done to prevent the call center workers from distracting each other from their work by milling around the coffee machine. At least, that was the general idea. But it was a bad idea. As the institution continued to analyze the most productive teams, the bank started to learn what distinguished them from the other, less productive workers: the seemingly unproductive coffee breaks that workers were spending together. The more the workers chatted with their coworkers in the break room, the faster and more efficiently they worked when they were back on the phones. The end solution was that coffee breaks were planned so that an entire team was always given a break at the same time (their calls were simply forwarded to a different team, which wasn't a problem for a large call center). Social interaction, which might present a distraction while working, was allowed to unleash its productive power during breaks. Back on the job after their common coffee break, the pace of the workers increased by 20 percent (measured by the number of calls processed). Reported customer satisfaction increased by 10 percent. When the bank finally rolled out this concept for all 25,000 call center employees, their combined productivity generated a total increase of fifteen million dollars.[20] That just goes to show what can happen if you understand what powerful effects distraction can have on the brain.

A mental diet

IF WE DON'T take the opportunity to mentally recharge during a break (preferably by talking to others), we will eventually be derailed and disrupted by distractions. The more homogenous the task, the stronger is our tendency to be distracted. The bad news is that our susceptibility to distraction increases over time. Someone who uses several different types of technology will get increasingly worse at toggling back and forth between their smartphone, paperwork, computer, and personal conversations. Haven't you ever noticed that the people who use their various devices the most are also those who seem to be the worst at prioritizing and the most susceptible to interruptions? Young people (from the ages of thirteen to twenty-four) who took concentration tests for a study proved to be particularly prone to distraction if they often used their smartphones for supposed multitasking (i.e., texting while watching a video).[21] Anyone who is constantly bouncing back and forth and getting distracted will eventually lose their competency at weighing incoming information and stimuli until it becomes a vicious cycle. The good news is that we are able to train our thalamus to be more resistant to smartphones and media distraction.

When does distraction have the most allure? The first time, when it suddenly pops up. Thus, the best thing to do is to avoid your first impulse—for example, looking at your phone right after it buzzes. Don't pretend to not pay attention to your smartphone, but instead, choose to actively ignore it. When it vibrates, register the buzzing noise and consciously decide not to reach for it. If you actively choose not to attend to your phone, you are also actively choosing not to follow the incoming stream of messages. This extremely valuable method of prioritization is picked up on by the thalamus, which learns it too and will, in time, grow ever more resistant to distraction.

The next step is equally important: regulating your use of technology. The most dangerous threat to our ability to concentrate is not that we use our smartphone during working hours, but that we use it too irregularly. By checking our emails every now and then on the computer and our text messages here and there on our phone with no particular schedule or rhythm in mind, the thalamus loses its ability to effectively filter. The solution is to regulate your devices as if you were on a strict diet. When it comes to nutrition, sticking to a fixed time plan for breakfast, lunch, and dinner allows your metabolism to adjust, thereby causing less hunger during the in-between phases. Your belly will start to rumble around 12:30 p.m. each day, but that's okay because that's a good time to eat lunch. If something unexpected happens, you can add a snack every now and then to get fresh energy but your metabolism will remain under control. It's the same with our brain when you put it on a regulated "media diet." Those who only check their messages or emails at certain times are less prone to binging on technology in between meals.

Inspiring distractions

IT'S A STRANGE thing how our brain filters are so robust and effective that we run into lampposts while looking down in deep concentration at our smartphones. But even the smallest things can distract us too. Why aren't we always concentrating equally? Why does it sometimes feel like pulling teeth to keep from getting distracted?

From a neurobiological perspective, distraction is not automatically something negative. Rather, it is important evidence that we are more than merely distraction-proof concentration machines. Someone with mental blinders on is only able to go in a single direction, possibly missing the turnoff to a better future. If we were unable to be interrupted, we would, of course, be resistant to distraction, but also to inspiration. We would be able to complete our given task with focus—but only this one single task.

Creative people allow themselves to get more easily distracted because their filter mechanisms do not work as well as those of other, less creative people. They are not as good at blocking out noise and often feel quickly irritated at banal background interference. This has even been measured in the brain. Especially creative people (artists, scientists, or designers) tend not to block out repetitive stimuli as much as less creative people. In other words, creative people have their brains organically adjusted not to get used to stimuli as quickly, to filter out less information, and to be more distracted.[22] For a creative person, a new noise is not immediately marked as an intrusion but has the potential to be something inspiring.

Many creative people therefore prefer to generate new ideas in environments that provide moderate distraction. For example, in a street café with a gentle stream of murmuring in the background, listening to subdued music, or undertaking mundane activities such as driving or going for walks. Interestingly, these kinds of activities, with moderate distraction, are particularly stimulating for fresh creativity. Not only were study participants especially good at coming up with original uses for everyday objects when they were surrounded by constant background noise (background noise from a cafeteria), but customers who were subjected to an inspiring sound (such as a steady stream of background conversations) also chose to buy more new, innovative products.[23] The optimal creative volume is seventy decibels, which is the equivalent of a single, rather loud conversation.

Whether or not we are creatively inspired or distracted has to do with the type of situation and our environment. Distractions are not always negative. If our goal is to come up with a new solution to a problem (such as developing a new slogan or design or a plan for the next kid's birthday party), light distractions can offer some inspiration. It's the amount that counts. People who are especially creative pay attention to the balance between concentration and distraction. They focus first on a problem, enter a zone of controlled distraction (for example, a street café), and then return again to a quieter place. This lets the brain optimally adjust its sensory filters, allowing, when needed, whatever is necessary to get through a task.

Speaking of getting through... I just noticed there are a few new messages in my inbox. If you'll pardon me for a few seconds...

8

MATHEMATICS

*Why the Brain Calculates Best
without Numbers*

ENOUGH WITH ALL the words, the long descriptions, and abstract ideas. We are going to turn now to a much more logical and universal method of describing the world— namely, the world of numbers.

No doubt half of my readers have just shrieked in panic because there is no discipline that seems to polarize people like mathematics. A representative Forsa survey from 2010 found that 40 percent of adults and 35 percent of students queried view math as their favorite subject, and 68 percent of those surveyed enjoy doing calculations in their everyday life. Unfortunately, those surveyed found it more difficult to have a sense for statistical correlations, with only 18 percent of respondents from the same survey believing that anyone else could think math is fun.

Really? Is it possible for anyone to have fun with numbers and calculations? I myself even think the previous paragraph had too many numbers and percentages in it. You end up stumbling through each sentence, tripping from number to number. What does something like "68 percent" even mean? You can't really feel the difference between 68 and 69 percent. And though calculations and individual problems such as 8 x 8, 27 + 59, or 3^4 are easy enough to solve, we quickly reach the limit of our skills. Even 3^5 starts to get complicated regardless of whether you are "good" at math, it was your favorite subject in school, or you are a professional mathematician. The truth is, our brain is bad at calculating. Okay, our brains naturally "calculate" in their own special way and work with their own rules to "calculate" information, events, or emotions. But the friendship sours as soon as numbers are thrown into the mix.

Even if you are a math expert, your skills fade enormously in comparison with a calculator. The calculator can figure out, within mere fractions of a second, the tangent of $\pi/3$ or the fourth root of 4.3^{34}. Even a prodigy would have to ponder this problem for a bit since our brain is not made for this kind of complicated arithmetic.

And yet, although even the cheapest calculators have been leaving math geniuses in the dust for decades, no one has ever thought to view these computational abilities as a threat to our mental supremacy. We do make such claims when it comes to successes in "artificial intelligence," even though the best computers are as dumb as a pocket calculator. Because though computers are able to calculate without error once the rules of math have been properly programmed into them, they fail to do anything at all with their computational results.

We, on the other hand, are able to interpret calculations and apply them to create an entire structure of thought. Computers compute (hence their name). Humans conduct mathematics. There's a difference. If you want to know what I mean, just open up a math book and read through the chapter titles: "Longitudinal vibrations of a beam,"[1] "Inverse of a matrix,"[2] or "Mapping of a half plane onto a circle."[3] This sounds like scientific poetry. It is rather marvelous to think of the mental deviations that our brain is capable of in mathematics. And yet, in a world that increasingly places more value on measurability, numbers, and spreadsheets, neurobiology must emphasize that math is not all that easy for our brains and, in fact, can be extremely cumbersome and prone to error.

Nevertheless, we are continually bombarded by numbers every day. There is no realm of life that can withstand being numerically quantified. But numbers are often only pseudo explanations, pretending to offer analytical clarity that our brains can hardly grasp. For example, what does a 40 percent chance of rain actually mean? Should I worry when a side effect of a medication causes a rash "in rare cases"? Or is it more dangerous to lie in the sun for ten minutes? Am I able to comprehend large numbers such as 285,000,000,000 or do I always have to draw on weird comparisons (about a twelfth of the U.S. federal government's annual tax revenue, or 1,080 times the population of Indonesia)? Does the average life span of eighty-one imply that someone else has to die nine years earlier if my grandma just turned ninety?

If you look closely, it becomes obvious that the brain does not do well with numbers (especially big numbers). And calculating numbers is not easy for us either. With a bit of practice or maybe with the help of some math trick or other, we might be able to solve a problem more rapidly. But even in that case, we don't stand a chance against a calculator app that's available for free on our smartphone. But don't fret. I am here to tell you that this is a good thing. Because the reason for this is probably the greatest strength of our brain—namely, we think in patterns, images, and correlations. We might be relatively bad at arranging number columns into tables, but we are masters when it comes to generating images and stories out of collections of numbers. Just ask an astronomer. When she prints out the data from her telescope, she principally has an endless row of numbers. No one can understand that. Only after the numbers have been arranged according to patterns do we start to get an idea of truly fantastic things like "red giants," "black holes," or "dark matter" in the universe. This is only possible because we are not very good at math. If we were, we would be just as unimaginative as a computer.

Numbers? No thanks!

WHEN WE CALCULATE numbers, probabilities, and large amounts, we often make mistakes. Even I am not immune to errors, though I am writing a scientific book. My publisher warned me: *Be careful not to write any formulas in your book! They won't be well received, will look too complicated, and will lose you readers!* That might be true, but I'm going to do it anyways because thanks to neurology we now know the reason why some mathematical formulas seem to be so unattractive, or even downright ugly, tending to scare off one's readers: they contain too many numbers and we don't like numbers. Would you like an example?

$$\frac{1}{\pi} = \frac{2\sqrt{2}}{9801} \sum_{k=0}^{\infty} \frac{(4k)!\,(1103 + 26390k)}{(k!)^4 396^{4k}}$$

Well, you might say, that really is pretty unwieldy. An equation to calculate a circumference of pi, featuring an almost chaotic arrangement of numbers. Awful. And you wouldn't be alone. It was chosen as one of the ugliest formulas by mathematicians.

How pleasantly refreshing, then, is this counterproposal, the most beautiful formula in the world according to a questionnaire given to mathematicians:

$$e^{i\pi} + 1 = 0$$

The math amateur will mockingly argue that this second formula is much shorter and therefore logistically more beautiful—simple and correct at the same time. In addition, there are no further numbers besides the one and the zero (and I have to agree though I am not a mathematician). But at the same time, this formula activates a region in the brain of mathematicians that is also switched on upon experiencing seemingly beautiful art or music. This region is known as the medial orbitofrontal cortex,[4] a part of the frontal lobe that is located directly under the glabella, the (usually) hairless area between your eyebrows. But this formula only activates that area in mathematicians. We mere mortals can hardly start to grasp such formulas and have no emotional ties to them, and our brains have thus not been found in studies to react in the same manner. For a good reason.

For someone with no real grasp of math, every formula looks equally complicated, regardless of whether there are a few or many numbers. Without previous education in math, humans do not possess an intuitive relationship to numbers or formulas. We cannot "feel" them in the same way that we feel the sun shining on our skin or the milky sweetness of ice cream when we eat it. We also can't feel a probability of 40 percent or 1 in 302,575,350. Mathematics is like a foreign language from another, abstract world that we are required, with much effort, to learn. It's similar to when you start to learn a "real" foreign language and have no feeling in the beginning for the words. We might not have any scruples about tossing around the worst swear words from other languages—French, German, or Spanish—but if we were to say the same words in English, we might do so a bit more quietly.

The reason is simple: humans are not number or language robots. We don't analyze words in order to put them into new formulas or to construct new sentences. This is what computers do, and they do not have a problem completing complex mathematical calculations for us. But we are really bad at remembering both vocabulary and formulas (see chapter 2). What we are good at is understanding things and putting vocabulary and formulas into context. The word "sun" is not just a collection of three letters that takes up forty bits of space on a hard drive, but rather it is a warming object that gives us good moods in the summer along with the occasional sunburn. Whatever image just flashed in your inner eye, it was certainly a feeling that you have around the idea of the sun.

And what do you feel about the number 9801? Nothing? Listen deeply to your heart, maybe there is something there... Or maybe not, since numbers are so abstract that we are seldom able to conjure stories, feelings, or images to correspond to them (with the exception of the number 13, about which some people report "feeling" a sensation of bad luck or misfortune).

Numbers, please!

SOMETIMES PEOPLE SAY that they think they are a "numbers person"—someone who is really good with numbers and at calculating. Now, numbers, in and of themselves, are completely inanimate, absolutely meaningless. And as important as numbers are in describing the world, they can only do just that: describe it. A number has never changed the world, although the stories *behind* the numbers certainly have.

Numbers aren't worth anything. When I say to you "Three!" you might say, "Three what? Three Musketeers? Three wishes on a magic lamp? Three Stooges? Three fingers on the right-hand rule?" And you would be justified in your questioning because without any context, the number three doesn't hold much meaning. However, your brain is still able to process the number three and can grasp that three is more than two.

Which leads to an interesting question: whether numbers exist apart from humans or whether they are a human "invention" used to organize the world. Mathematics is a subject of the humanities but does that mean that it would no longer exist if there were no one around to think about it? What happens to "three" when no one is left to think about "three"? I promise I am not about to launch into a philosophical discourse on epistemology in this book, but from a neuroscientific perspective there are a few exciting and very clear indications that numbers are more than artificial crutches used by humans to organize the world. We actually seem to detect numbers in much the same way as we do with other things in our surroundings.

When we arrive in the world, we are not yet able to count, at least not in the traditional sense. Expectant parents can play as much classical music as they want for their fetus while it is forming in the womb, but the concept of mathematics can only develop gradually, step by step, after birth. Nonetheless, as humans we already possess all of the anatomical prerequisites necessary for dealing with numbers. This is, to a certain extent, a mathematical basis that allows us to start thinking numerically. We are apparently equipped with three basic techniques to help us comprehend numbers from the very beginning.

Our basic mathematical equipment

ONE OF THESE basic techniques has been called "subitizing" in science, which means something equivalent to "instant counting." For example, how many points do you see here:

No problem, there were four points. And how many do you see here:

This is much harder since we can only immediately grasp the number of a small handful of objects. When we do this, we are not counting the objects, but rather we are directly "seeing" how many objects there are. This is a pretty good start for humans who need to use small numbers to orient themselves in the world.

A second basic number principle is our ability to estimate numbers. In contrast to the above subitizing, we cannot immediately grasp the correct number but instead are able to relate quantities to one another. Four is bigger than two. Ten is bigger than five. Further examination of number estimation shows us another reason why we are so bad at dealing with numbers. Imagine that you are tasked with organizing various numbers on a number line relative to one another—let's say the numbers 1, 2, 3, 5, 10, and 50. Anyone who paid attention in primary school will remember that this number line would look something like this:

123_5_____10_____ 50

But if you ask second graders or adult members of certain Indigenous groups who do not often have to count above five, you would get something like the following line:

1__2____3_____5_____ 10_____ 50

The bigger the numbers, the less precise is the estimation. This is because our ground-up comprehension of numbers is not linear but rather logarithmic.[5] In the end, it isn't the absolute values of numbers that interests us. We get much more excited about the differences and comparisons. You know this from evenings when you dip into a bag of potato chips. After munching delicately on the first chip, the question arises: "Should I keep eating? If I eat a second chip, I will have eaten twice as many as I have now." This second potato chip might require a fair bit of persuasion to overcome your guilty conscience. But by the time you've reached chip number thirty-nine the dam has already broken. It no longer matters whether you've chomped down forty or forty-one chips. All you know is it's a whole lot of calories.

This phenomenon is known in science as Weber's law: the subjective perception of sensory impressions follows a logarithmic dependence. Let me offer a practical example. If we wish to be able to detect a difference in weight, the additional weight must comprise at least about 2 percent of the total mass. Say you put your shopping basket, weighing 15 pounds, onto the scale. If you add a chocolate bar weighing 3.5 ounces, you will hardly notice the difference because it is only when something weighs 5 ounces (which is about 2 percent of the original mass) that it really becomes noticeable. The Weber law is irritatingly impractical if you are hoping to lose some weight. If you weigh 260 pounds, you have to lose over 5 pounds before you will perceive any change. Whereas someone who only weighs 110 pounds feels noticeably lighter after simply going to the bathroom in the morning.

The Weber law also applies to our understanding of numbers, but only when it comes to definite numbers (see: potato chips). Once these form a pattern, we are much better able to value and compare them.[6] This is what makes it possible for us to relate a totality of individual parts to another totality without having to count each individual piece precisely. For example, if you spread out two different types of potato chips in front of you, the red barbecue-flavored chips and the yellowish onion-flavored ones, it is very difficult to tell at a glance how many barbecue chips you can see or whether there are more barbecue than onion chips. But if you crumble up a bag of onion chips and a bag of barbecue chips and lay them out side by side, it will be much easier for you to see from the crumbs which pile is larger. Anyone with small children at home knows just how easy it is to replicate this experiment.

This third basic technique equips the brain early on to be able to calculate more precisely in the future. And indeed, this mathematical comprehension seems to be innate, as even children who still have not developed language skills or Indigenous groups who only use a one-two-many method of counting are able to apply this principle. In addition, the very fact that our numerical estimation follows Weber's law indicates that we are able to perceive numbers rather than that we create them artificially. And just like any other kind of sensory input, we are able to become familiar with numbers. Study participants who were initially shown an arrangement of 30 dots were able, later on, to correctly estimate another group of 30 dots as consisting of 30 dots.[7] However, if they were first shown a picture of 400 dots, they incorrectly estimated the number of dots in a 100-dot arrangement as consisting of 30 dots. It's a little bit like slowing down suddenly on the highway. Your rapid decrease to 35 miles per hour feels like only 15 miles per hour. Numbers are thus not an invention but rather something we perceive much as we do other sensory input. It's as though they exist apart from us. This impression is further strengthened if we take a look at the way our neurons process numbers.

Democratic votes by the number

WHAT HAPPENS IN your brain when you see the number 5? Or five objects at the same time? Or one after the other? Or hear five tones in a row? Interestingly, the same cells always react to this numerical stimulus regardless of whether it comes via your vision or hearing. This is because we have special number neurons that are responsible for concentrating precisely on this concrete form of perception.[8]

In humans, these neurons are located in two different regions of the brain, the posterior parietal lobe and lateral frontal lobe of the cerebral cortex. Both regions are perfectly positioned to represent abstract numbers since they receive preprocessed inputs from the hearing and seeing centers of the brain. It doesn't make any difference to us whether we see five dots or hear five peeps. In both cases we generate the abstract construct "five."

Our numerical comprehension begins in the posterior parietal lobe where the first number neurons associate sensory input with a number. This information is then passed on to the lateral part of the frontal lobe where it is further abstracted and, if necessary, calculated against other numbers. We apparently also have specialized computational neurons that are able to code individual arithmetic operations (for example, "more than"). In other words, we are able to perceive the numerical property of objects in our environment (other species can do this too), but then have the capability to further abstract these properties (which primates can also do) and finally to construct a complete mental calculus from it (this is where even the apes fall behind).

Numbers have a pretty good reputation in our world because they come across as objective and factual. A five is a 5, is a V, is a ̶H̶H̶, no matter where you are in the world. But in the brain, this process is much more democratic. That is, the number that we perceive is, well, "voted on" every single time we perceive it, implying that numbers are principally anything but objective in our brains. Numbers are more of a majority decision.

Picture this: you are inside of the intraparietal sulcus of the posterior parietal lobe. Around you are the best of the best—namely, neurons that are specialized in all of the different numbers. There are some that become especially active when the number 3 is presented and others that react strongly to the number 6. But our neurons are not perfect. They are much more prone to error than a computer component. Thus the number-3 neuron also reacts somewhat to the neighboring numbers 2 and 4 and perhaps even a tiny bit to the numbers 1 and 5. The same is true of the number-6 neuron, which shakes its hips slightly at the sight of numbers 5 and 7. An individual neuron is never completely certain if it has been stimulated by its own personal number. But for the brain, it is very important to be able to recognize a particular number. All of the neurons gather together in the parietal lobe to take a "vote," each of them carrying their particular number profile. The sum of the total activity produces a result that indicates which number we must be talking about. If the 3 is presented, there will be more number-3 neurons present firing off more powerfully than, say, a number-8 neuron or a number-1 neuron. The result is guided through the lateral frontal lobe. The neurons there use the number for a calculation or for making a decision (see chapter 9).

This numerical thinking has two benefits in practice. Firstly, it allows us to recognize differences in surroundings quite rapidly, a phenomenon known as distance effect. This holds that the greater the distance between two numbers, the easier it is to compare them. It is easier, for example, to distinguish 2 from 10 than it is to distinguish 5 and 6. This effect is beneficial if you don't get bogged down in details at first glance but rather need to do a quick and dirty evaluation of your surroundings. The reason for this lies in the somewhat "blurred" activation of the number neurons. The less the adjacent number neurons overlap in their activity (i.e., the number 5 and the number 6), the clearer the signal will be. If your objective is thus to fake someone's balance sheets, you are better off switching 11,100 with 11,110, which is going to be hard for someone to catch, whereas replacing 11,100 with 45,879 could be considered a rather risky move.

The second practical benefit is that small numbers are easier to process than large ones. Because the activity spectrum of number-neuron groups in our parietal lobe is constantly widening as larger numbers are added, it becomes increasingly difficult for us to distinguish these numbers. We shouldn't worry too much about this because numbers get so big eventually that the difference between them hardly matters. Whether Manchester City becomes English soccer's Premier League champion by 70 or 71 points is not that important. But whether a lesser team has 38 or 39 points by the end of the season has the potential to make a huge difference in their overall placement. We are especially able to grasp smaller numbers that can be counted on our fingers. But at some point, as the numbers get bigger and bigger, we start to lose our reference point. Maybe this is why the Mundurukú people from the Amazon only count up to five. Which doesn't, by the way, hinder them from correctly comparing larger numbers with one another.[9]

Patterns, not numbers

WE HAVE IN our possession some very basic equipment that allows us to quickly estimate small dimensions, to draw comparisons, and to count small quantities. And that's about it. Our arithmetic capabilities are very limited. Why should our brain have developed a system that can process numbers beyond 10 to the 11th power? Isn't it obvious enough that entire states are that much in debt or that Silicon Valley companies have that much of their cash in stocks? Our brains check out entirely when it comes to probabilities, and our neurons don't do much to help us calculate roots. In fact, we can do little more than the four basic arithmetic operations and count a couple of dots, and yet we have nonetheless been able to develop topological spaces, describe Noetherian rings, and calculate Apollonian circles. Why? Because we are able to do what computers cannot: expand the rules of math and apply them in innovative new ways. Let's leave the boring old, number-fixated arithmetic operations to the computers. Anyone who boasts about being a whiz at calculating tables efficiently, quickly, and error-free is basically putting themselves on the level of an algorithm. In which case they can't really complain if subsequent software is soon developed that rapidly trumps and replaces their abilities. Working efficiently with numbers is not one of our brain's strengths. On the contrary. At the same time, there are other things that computers won't be able to replace for the foreseeable future, and these are processed by the brain in the numerical regions of the cerebrum just mentioned.

One might say that math and language are interconnected. That without language, there would be no sense of numbers.

Some say that whatever we are able to name is what exists in the brain. Following this logic, the Amazonian Indians would not be able to add 12 to 34. But this isn't true. Language has almost nothing to do with math in the brain.

Consider the following sentence: "There exist nondiscrete spaces whose connected components are reduced to one point." Ah, I see, you might say, that sounds logical enough. And what about this: "Following the Vatican, Monaco is the smallest country in the world"? Sure, that sounds about right too. Both sentences are quite complicated and require the reader to create relationships, comparisons, and abstractions, and in both sentences, language plays an important role. And yet the first sentence is not processed in the language centers of the brain (at least not if one is a professional mathematician), but the second one is.[10] But if one is not well versed in higher mathematics, both sentences are processed in the language centers of the brain. Mathematicians, on the other hand, employ the same brain regions normally required for rudimentary numerical reasoning to dissect these complex mathematical claims—even though they don't contain a single number. At the same time, the regions of the brain responsible for facial recognition are less active among mathematicians, which perhaps points to the widespread prejudice about "sociophobic" math nerds sitting alone in their rooms calculating formulas. I would personally like to say I don't agree with this stereotype and would now like to take a moment to bestow on mathematics the neurobiological significance it deserves. If nothing else, because mathematical thinking is the first-class proof that we are much more than mere biological calculating machines. In truth, we are the very opposite.

Misused brain regions

THE BRAIN'S WEAKNESS for complicated calculations or big numbers turns out to be one of its greatest strengths. Because it is only in this way that we don't get bogged down in a jumble of numbers but are instead free to interpret them. Mathematicians don't think in numbers but in patterns and pictures, in relationships and spaces. Brain scanners clearly show mathematicians activating areas responsible for image processing and pattern recognition.[11] Numbers are and remain uninteresting. But numbers in correlation and their dynamics when calculated are much more exciting.

It's a bit like playing chess. What is the professional chess player looking at when they gaze down at a chessboard? One might think they are looking at the game pieces, which they have to move. And they are. At least, they are looking at the piece most relevant for the next move. But even more important than the chess pieces are the open spaces around them, and it is to these that chess players pay particular attention.[12] It is only by considering the spaces in between and using them in a meaningful way that new patterns can emerge.

We make sense of our world in the same way, whether we are a mathematician, a chess player, or a truck driver. What's important is our ability to recognize relationships. This is what our brain is equipped for, though it also relies on the brain regions responsible for basic math and numbers. We come into the world with a sense for small numbers (just as we come into the world with a sense for time and space), regardless of whether we are born in the Amazon or in San Francisco. But it is only through education that we are able to employ the rudimentary thinking tools required for abstract thought. Evolution certainly has not equipped our brains with the innate ability to diagonalize matrices. This doesn't matter much, however, because our brain's basic math regions let themselves be misused for these purposes. Experts call this phenomenon preadaptation, which is basically to say alternative applications of our abilities. We do this all the time; for example, once we've mastered the skill of building a fire, we can also learn to cook our soup or provide a spark for our sports car's internal combustion engine so it can release the 310 horsepower from within. In the same way, the brain regions responsible for perceiving and processing numbers can also help us develop and apply abstract mathematical pictures, formulas, and concepts.

This is only remotely related to numbers. We do not think in numbers or codes but rather in patterns and images. In any case, this is the evidence unearthed by experiments in which mathematicians process mathematical sentences, not with their language centers, but with their numerical brain regions and draw on image processing areas during abstract thinking. Thus, when the Austrian philosopher Ludwig Wittgenstein said, "The limits of my language mean the limits of my world," he was not correct. Because where language ends, the world of the brain begins. We are more than biological machines processing numbers, symbols, and letters. It's only when we are able to create images from these components that they take on emotional significance for us.

This doesn't only apply to math; it's true of every form of human thought. Our brains get excited only when we emerge from the lowlands of numbers and symbols and begin to construct stories. Even as I write this book, I am not thinking about the letters and words that I am typing on the keyboard but rather about an image, a message, an important thread that I wish to get across to you. And you are not at all interested in the individual letters or numbers in this book. However, you are able to formulate an exciting idea in your mind: that our brain is necessarily error prone when it comes to numbers and letters in order for it to free up space for us to think abstractly. Writing a text with ideas in the back of one's mind is fun. Proofreading the text once it's written, on the other hand, is not. I would like to give a shout out to the very skilled copyeditors of this book. I only have the greatest respect for your orthographic achievements—since our brain isn't actually wired to do this kind of work.

Simply writing a book is a strategic error. We all know that a picture is worth a thousand words. A ninety-minute film with twenty-four pictures per second is, by this reasoning, able to say more than 129 million words put together. And if any of my readers might be considering making the contents of this book into a scientific Netflix series, the total of 90,000 words in this book would take about 3.75 seconds to convey in images. That books can be sold at all is due to the fact that our brain is not as interessted in the exact text as much as it is in images the book conveys. Perhaps that's why you may have failed to notice the spelling mistake in the previous sentence. Don't worry! You still managed to form a thought in your brain, and that's what really matters.

The language of the brain

THE BRAIN IS relatively bad in arithmetic but is able to appropriate its numerical regions to construct abstract thought. Mathematics is practical proof that we are much more than accurate calculating machines since machines are quickly able to find the total of 145,099 + 27,845 but are unable to do anything with it.

The fact remains: the language of the brain is full of patterns and emotions. This has practical effects on our behavior because no matter how much you might try to impress others with your numerical facts, you don't stand a chance against a picture. This is how Apple sells its iPhones. Not by flaunting all the technical details of its hardware (by the way, no one really knows how fast the iPhone processor really is) but by generating the image that if you have an iPhone, you're hip, you're connected with your friends, you're snapping photos of your relatives laughing, and you are able to share ideas with the world. It doesn't really matter how fast the dumb processor is.

Imagine you have two different options for donating money to a children's aid foundation in Africa. In the first option, you read that your money will reach a seven-year-old girl named Rokia who lives below the poverty line and who, along with her family, often experiences hunger. Your donation would significantly improve Rokia's life. In the second option, you read that due to heavy rainfall, the rice production in Zambia has decreased by 42 percent, and there are now three million citizens suffering from hunger. To which cause would you rather donate? If you are like the participants in a study from 2007, you would choose the first case. If one puts a photograph of wide-eyed, sad Rokia next to the written material, participants will donate exactly twice as much as those who only get to read a few numbers and statistics.[13] This is true, even though the second group of participants read about the suffering of a much larger group of people, and not only about one individual little girl. But this information was unfortunately abstract and numerical and thus less heartwarming.

We come into the world with a sense for small numbers and amounts but, unfortunately, not for probabilities and statistics. Our modern ability to compute has helped us to master these emotionless percentages that make us believe we can better understand the world, when in fact we forget that every percentage must be painstakingly translated into an image in our brain. And we are very selfish when it comes to this, because our brain is not interested in what is happening to a statistical population, but only in what is happening to ourselves.

Would you bring an umbrella with you if your weather app predicts a 10 percent chance of rain? What this prediction means is that of ten different weather models for a particular period of time, one of them has predicted rain with absolute certainty. If you are unlucky, this is precisely the weather model that is going to find you in the exact moment when you venture outside. It doesn't help that the exception occurs so rarely—unless, of course, you are the exception.

Because we have so little sense for probabilities, we are always falling into the trap. For example, we voluntarily pay for insurance. Statistically speaking, this is sheer tomfoolery because we end up paying more for insurance against a fire than we statistically would need to pay for damages due to fire. If our house is worth a million dollars and the probability of a fire with a total loss in the coming year is 1 percent, the total expected damages would cost $10,000 (loss times the probability of occurrence). But you always end up paying more than $10,000 insurance over the years, otherwise the insurance system would never make any money. This brain weakness is precisely how insurance companies earn the big bucks. It's only small consolation then that—here in Germany at least—insurance companies fall for their own sales tricks and insure their own losses with reinsurers. (And in case you didn't know, we Germans love it! We adore our many insurance policies. No one has more irrational angst than Germans do.) But really, insurance companies should know better.

"So?" you might say in response. "What if I end up being the statistical exception? In that case, statistics won't help me one bit." And you would be right because probabilities and numbers are not tangible. But stories are. And this is why the same people who insure themselves against a volcanic eruption in Northern Germany are also those who buy lottery tickets to seize their one in 140,000,000 chance to get rich.

Remain objective, think like you're crazy

AS YOU CAN see, our sense of numbers misdirects us to the strangest behaviors, especially when we abandon our basic "hereditary" number abilities and start to deal with very large or very small numbers or probabilities. The more far-reaching the decision that we are trying to make, the more we should try to avoid falling for individual exceptions or fates. Don't forget that our brain, which is helplessly bombarded by intangible percentages in our world, is doing everything it can to make sense of these numbers by translating them into meaningful images. So, before the next time you allow yourself to be blinded by an emotional story, pause and consider that the images generated by your own brain might sometimes trick you. Your brain sometimes doesn't realize that it is narrowing its image too much because the benefits of this kind of thinking are generally much greater than the disadvantages.

We are ultimately able to rise above the world of numbers and data. Computers might be able to gather, correlate, and combine gigantic amounts of data quickly and flawlessly, but they are not really equipped to interpret it. Only we are able to assign a "value" to numbers or to give meaning to our world. Much more important than the original numerical property of a number is its unit. Five apples are different than five city blocks in Manhattan. Both times, the "five" is processed by the same neurons in the numeric region of the brain. But it is only when these react with the other brain regions that we start to form a picture, a pattern. And only then do we have what we need to change the world.

9

DECISIONS

*Why We Risk Too Much—Yet
Still Choose Wisely*

IMAGINE YOU ARE participating in the game show *Who Wants
to Be a Millionaire?* Since I'm fairly certain that my readers
are mostly designated knowledge experts, let's say you make
it to the $500,000 question. You feel like you know the answer
to the question, but you aren't quite certain. What would bother
you more: to choose the wrong answer and drop back down
to $16,000 or to pocket $125,000 and walk away, finding out
that you actually did know the correct answer? You have the
choice of either losing $109,000 or not winning $375,000.
What hurts more?

From a mathematical standpoint, the case is clear: even if
you don't have a clue about the right answer and would just be

guessing blindly, you should absolutely take the risk! Because the anticipated value of a gain of $375,000 (with a 25 percent chance of correctly answering the $500,000 question) is higher than the anticipated value of the loss of $109,000 (with a 75 percent chance). A ratio greater than one. You definitely have to choose to guess, regardless of the cost. You will never face better odds in your life, not even if you put everything on red in roulette.

In the previous chapter, you learned that the brain is pretty bad at dealing with numbers. To be more precise, it loathes abstract numerical structures and cannot grasp them emotionally. A statistical anticipated value greater than one? This won't be much help to you if you do take the gamble and end up like a mathematically opinionated loser. This is why you see that in the decision-making process, such as in the $500,000 game-show question, people tend to have more anxiety about losing what they have gained than losing what they might win. This is a built-in mechanism that we have (where exactly this is built in we will see shortly) to protect us from too much harm.

Of course, this can also backfire. Just ask Ronald Wayne, the third founder of Apple along with Steve Jobs and Stephen Wozniak. A week after cofounding the company, he got cold feet and sold his share of the company back to Jobs and Wozniak— for $2,300.[1] If he had kept his stocks, he would be a billionaire today instead of spending thirty dollars a week to play the one-armed bandit machine in a Nevada casino.[2]

We get cold feet. Clearly. But this doesn't keep us from making risky decisions throughout our lives. We drive over the speed limit on the freeway, spend over 75 billion dollars a year in U.S. casinos, and love eating raw cookie dough. Sometimes we take deliberate risks and whiz down steep slopes on narrow boards, buy Tesla stocks, or undertake the biggest risk of all—and get married. With such a lifelong, all-in investment with a 50 percent probable default rate, an unknown rate of return, and no opportunity of distributing the risk, any investment advisor would put their head in their hands at such a decision. And yet almost five million people in the U.S. and Canada get married each year. What are they thinking?

Our decision-making process is clearly irrational. Sometimes driven by anxiety and a need for security, at other times it is motivated by a longing for thrills and adventure. And we apparently don't rely on factual considerations but rather reach our decisions impulsively, intuitively, and instinctively. Our rational cerebrum doesn't seem to have much say in the matter. Or does it?

The decision-making processes in our brain are in no way incomprehensible. The fact that we so often take risks or abandon our better intuition is only a superficial disadvantage because it seems to contradict our ideal of what the "right" decision should look like: that it should be well founded, factual, risk-reducing, and sustainable. And preferably not ruled by our gut emotions. This is not always the case, though, because our lives are not a series of probabilities. Our brain is prepared for this, however. It possesses a clever decision-making system, which though it may sometimes lead us unnecessarily into risk, gives us a decisive informational advantage.

Decisions are not calculations

BEFORE WE TURN to focus on our brain's totally clever decision-making system, I'll give you one more important idea to clearly illustrate that decisions are no trivial matter—and that our brain is pretty good at decision-making. Or, at least, that it is better than any computer in the world. It is in the nature of decision-making to have a moment of uncertainty. If you are looking for an absolutely foolproof, safe, and logically justifiable decision free of risk, you can forget about it. Because it's a no-brainer—literally.

Here's an example. The question "What is 87 x 24?" is not answered by making a decision. Whereas the question "Should I get married?" certainly is. If I am calculating something that is completely objective, I am not in the process of decision-making, but rather I am solving a problem which any computer could also do much more masterfully than I. Just give a computer enough data and a reliable formula, and it will spit out a result. And if the problem gets complicated, it can appear as though a computer actually did make a decision (think about autopilot in airplanes or a lane guard system in a car). But this is not the case. In reality, the computer merely solved a complex computational problem. Making an autonomous decision, on the other hand, requires something more. An algorithm essentially chooses one option among many, according to a particular set of rules.

For us humans, however, deciding is not about following a predetermined control scheme. Rather, we are able to piece together our own personal tools for decision-making out of many different rules, each and every time. The decision to get married is not the solution to a numerical puzzle because numerical rules are anything but unique. Perhaps you are certain you want to get married in a church? Or have an open-air wedding? Or maybe you have your heart set on a bachelor/bachelorette party? Or you need a five-year trial period first? Or you have to first explain your decision to your parents and siblings? Or else you have to play chess three times with your partner before you'll really know? Every single question might be important, but just how important is something you determine over and over again.

If the lane guard system in a car actively steers and makes certain that a car stays in its own lane, this is not an example of the car making a decision. This happens because someone activated the lane guard system. In the future we may see completely independent, self-driving cars, but even these vehicles won't be able to make decisions in the same way that humans do. Real decisions involve more than mere arithmetic and rule following. They are much more about the creative application, interpretation, and rebalancing of rules. They are subjective and never completely predictable. If they were, we would not be free.

In other words: there's no such thing as an objective, calculated decision. An image recognition software program may be able to "view" a photo and determine with 99 percent certainty that it is a face, but one should be careful before claiming "the software decided that it is a face." Actually, what the software did was solve a numeric task. We are always hearing about how "computers decided for us" or that they "take over some of our decisions." But from a neuropsychological perspective, these are not really decisions but rather simple input-output relationships. If you provide a computer with enough data, it will calculate the most likely, the best, or the cheapest option from all of the various possibilities. But in each of these cases, it is already clear which criteria should be used to rate the final choice.

Our decisions are also rated. We are punished or rewarded, by ourselves and by others, though we don't know what is going to happen before we make the decision, and therefore the criteria for rating our decisions are not always fixed. When is a marriage deemed successful? After at least seven years together? Once you've had two kids? Or when you're finally able to maximize your joint tax credit? Our brain does not simply rattle off a standardized decision-making program. Instead, it has developed a dynamic and flexible (and quite individualized) system with which we would, under totally different circumstances, reach a very different decision. The thing is, we may well be making the wrong decision, but there's simply no way of knowing this until we've tried it. This is the nature of decision-making, and it's not a bad thing. It isn't as important to the brain that we make the best decision as it is to make a decision that we are able to own up to later on. This is because we are able to influence our decision-making *process* completely, while we are often able to influence the *result* only somewhat. Get married, and I guarantee you'll find out what that means in practical terms.

Decisiveness is upside down

FOR THOSE WHO have always suspected that the brain decides not rationally but emotionally, the most recent neurological findings that back this up might not come as much of a surprise. In short, there is really no decision that can be based on facts alone.

Our brain networks in a very complex way and even supposedly simple actions such as language, image processing, or hand movements require the interaction of numerous areas of the brain. When it comes to making and implementing a decision, almost every brain region might be called upon, which poses a bit of a challenge for neuroscientific research. There is no "decision-making region" of the brain, no boss telling everyone what to do. Rather, decisions are formed in our brain "from the bottom up," starting in the emotional brain regions, spreading to the rational regions, and finally being translated into concrete actions. This threefold decision-making process is described in science as the "affect-integration-motivation framework."[3]

Step 1: The affect determines the direction.

Before anything gets decided in the brain, it first has to decide in which direction it is headed. And the most important question regarding direction is: Should one behave so as to receive a reward—or to avoid punishment? In order to measure its award/punishment goals, the brain has developed two separate nerve connections. They are located where the spinal cord ends and the brain begins, in the midbrain area at our neck. In contrast to its meaning-laden name, the midbrain is only 0.59 inches long and is a significant switchboard for bodily reflexes such as the control of our breath or our gag reflex. I would like, at this point, to once again refute the idea that we are able to make rational decisions. Because we have, right at the center of our midbrain, these two nerve connections that are either seeking reward or avoiding punishment. The "reward nerve" extends into the limbic system, or more precisely, into the nucleus accumbens. The other nerve connection stretches from the midbrain into the insular cortex. The insula does not seem to be very important to our concept of time (see chapter 5), but it does generate an unpleasant feeling of anticipated punishment, a negative sense of anticipation. This affect system determines the general direction. For example, the $500,000 question: I could either look forward to winning much more money (nucleus accumbens active), or I could wish to avoid losing everything with one false answer (frontal insular cortex active).

Step 2: Integrate feelings and facts.

We are much more than purely affective beings mindlessly chasing after reward. Although... essentially we are. Even though we rationally weigh the matter at hand and turn to our plans, our goals, and our knowledge in order to reach a decision, we do so primarily to validate our emotional impulse. First comes the goal, and then come the reasons and explanations that we seek to explain why we should pursue that goal. This is exactly what takes place in the anterior cerebral cortex (if you really want to know, it's located in the medial temporal and the lateral frontal cortex) that is directly behind our forehead. Our impulsive midbrain not only sends its emotional signals into our limbic system, it also shoots them toward the same anterior cortex. It is very widely networked and also receives connections from the cortex, the diencephalon, and the limbic system. In this way it integrates our memory, values, and inner state with our basic emotions. This is how our surrounding environment may influence our decision-making process, something that specialists call "framing." For example, if we are facing the $500,000 game-show question and have already taken mental possession of the amount of money needed to buy a 300-square-foot apartment in San Francisco, we are more likely to risk answering the question than if we are sitting directly across from another contestant who just fell back to $500 by answering incorrectly. Our frontal lobe is the important hub for these inner and outer influences. It passes its assessment back to the nucleus accumbens. The whole thing is like a game of Ping-Pong: emotional feedback is sent from the limbic system to the frontal lobe where it is compared with our plans and experiences and then sent back. This game between emotional and rational brain regions repeats subconsciously about three times until our decision-making goal emerges more clearly.

Step 3: Motivation to action.

The final step toward action is the logical consequence of this process of weighing and evaluating. In neuroscience, we call this prompt toward movement a *motivation*. The result of our decision-making process is sent as a particular activity pattern to the movement regions of our cerebral cortex (located directly under the top of our skull). Of course, the concrete movement patterns still have to be worked out at this point, but why else do we have a cerebellum that is able to carry out this bothersome task?

The gambling criterion

THIS IS QUITE a well-balanced system. Very early on, the brain has a feeling of where it would like to go and subsequently creates a few arguments that can back up its original emotional decision. First come the emotions, then comes the conscious mind power. This is helpful if we have to decide something in an incomprehensible situation. And yet we are still prone to choosing falsely because our decision-making system has three weaknesses: it is afraid of losing; it allows itself to be influenced by other risk-taking people; and it is simply too curious.

Let's look first at our fear of losing. When scientists study risky behavior in the lab, we usually employ so-called "bandit tasks" because they resemble the conduct of a one-armed bandit: participants are given some money and may then decide whether to gamble or to take home the money. Every participant, of course, wants to play for as much money as possible. However, our fear of losing also plays tricks on us. If, for example, participants were given fifty pounds (about US$65—this study was conducted in Britain where folks have a particular fondness for gambling), the decision to gamble was influenced by how apparent one's chances of loss were. If the researcher told participants that in the worst-case scenario they would go home with twenty pounds, only 43 percent decided to gamble. But if they were told that they could lose thirty pounds by playing, almost 62 percent decided to gamble.[4]

This is pretty remarkable since in both cases the researchers principally said the same thing, just from two different perspectives. On the other hand, the risk of loss certainly does shift our assessment of risk. Brain scans have shown that the regions of the brain that (among others) process loss assessment and punishment (the amygdala and the anterior insular cortex) are especially active when under threatening conditions such as the above scenario. Still, we don't take high risks and gamble because we are afraid of losing—we also want to be paid for taking a gamble. Namely, a possible win must outweigh the expected loss before our risk assessment will permit us to start playing the game.[5]

What this means in practice is the type of wisdom you see scrawled across clever motivational posters: "Those who try might lose, but those who don't try have lost already." This is, of course, total nonsense. What it should say is: "Those who don't try won't win anything." But this doesn't grab us by the throat in quite the same way because it does not cater to our fear of losing.

From fear to risk

IF YOU WERE to look at the way we behave, you might think that we actively pursue risk. At the end of the day, our life is any-thing but one hundred percent certain, and we are constantly running into danger. Every year, over 10,000 people are killed by alcohol-induced auto accidents in the United States. Twenty million people in the U.S. contract a sexually transmitted illness because they were too careless, and around twenty-seven mil-lion people worldwide get addicted to drugs because they were looking for a kick. We watch action films on TV, heckle quiz-show participants to take risks, and cheer on our favorite sports teams to play aggressive offence. All of the time our brain has one goal in mind: not to lose.

We have just looked at one reason for risky behavior—namely, the fear of losing. This is a paradox, but when we are worried about a potential loss, we are quick to enter into risk in order to avoid the loss or hurt. A lot of people think it is our greed that drives us to take risks. And this is not wrong because this also motivates us, like fear, into risk. When are box-ers the most dangerous, for instance? When they've just been hit. When does the German national soccer team risk playing offensively? When France is up 0–2. When do we drive like maniacs on the freeway? When we think we might be late.

Management consultants know how to use this fear prac-tically. If you wish to initiate a change within a company, you face difficulties from the very beginning because no one wants to challenge the status quo. You could try to illustrate for the employees how great the company is going to be after the change or paint a rosy picture of the future. But this won't generate the right amount of motivation. It's more effective to present a threatening scenario: "If you don't change, in five years the company will no longer exist in the marketplace." It's only when the end of the rope has been reached that people will decide to enter into risk.

In the same way, people's decisions can be influenced by such messages, as the media regularly demonstrates. Our society is aging, the media tells us. Skilled workers are in short supply, so we have to change our educational system. New forms of digital media are making us all dumber; we have to read more books. (I am all for this!) The disruptive Silicon Valley start-up culture is going to devastate technical innovation around the rest of the world, unless we make ourselves as hip as these Californian up-starts. Whether or not these individual claims are true doesn't matter here. The point is that this method of publicly broadcasting such opinions seems to be working. It is working because there is nothing quite as contagious as fear. Nothing, perhaps, except greed.

The signature trait of stock market success

REASON NUMBER TWO for risky behavior is thus something that we "catch" from others. "The mob is swayed, uncertain in its mind, then wherever the stream flows, flows behind," wrote Goethe in *Faust*. And two hundred years later, thanks to neuroscience, we know where this uncertain mind is located in our brain: in the caudate nucleus.

The name sounds somewhat less poetic than my lyrical Frankfurt colleague expressed it in the nineteenth century. However, it does allow us to explain why we are more willing to take risks ourselves when we see others around us doing so. If those in a study are allowed to observe other participants risk-ing money, they are more easily drawn into a gambling situation themselves because their caudate nucleus becomes especially active.[6] This collection of neurons in the middle of our brain functions as the intersection between one's own behavior and the recognition of the behavior of others. The more closely these neurons are knitted to the frontal lobe region responsible for conscious thought, the more our decisions are influenced by the risk-taking behaviors of others, and the more easily we allow ourselves to be drawn into a similar behavior.

If we allow ourselves to be too conscious of the behaviors of others, we may experience short-term extremes. This illustrates itself in markets as bubbles, in art or music as hype, or in fash-ion as a flash-in-the-pan new style. In each of these cases, the finely balanced decision-making system in our brain shifts in favor of our more emotional centers. This has also been shown in brain scans through a technique called hyperscanning (which is not unlike the latest fashion trends).

Using this technique, not only can we measure the brain activity of individuals, but we can also study multiple indi-viduals at the same time. For example, two individuals might play a game together or barter, as they did in a study in 2014. Researchers measured the brain activity of participants while they simulated a market scenario and negotiated bonds. The researchers expected some of the simulation games to lead to an overheated market. But the reality showed that in every case, let me repeat that, *in every single case*, the market reached an overvaluation of shares ending with a sharp price slump. Researchers were interested in studying whether the brains of the successful market players (who sold their shares just before the peak) differed from the losers. And indeed, something emerged that seemed to be the "signature trait of success"—those who exited the market just at the right moment and made gains showed activity in their anterior insular cortex directly before making their decision. You might recall that this is the region of the brain involved in anticipating loss. If, on the other hand, one took a risk at this juncture, the nucleus accumbens (our reward center) was more actively involved in the decision-making process.[7]

You should remember that all of the participants found themselves in a volatile market, but only one group made the right decisions and got out in time. Not because they made a rational and conscious decision, but because, at the right time, they activated their emotional brain regions, which are aimed at risk mitigation and safety. As if they had an early-warning system in place to keep their greed in check. It is no easy feat to opt out of a quickly rising market. You will kick yourself if you watch it continue to rise after you've sold all your shares. But the brains of those who made the best decisions are characterized by being cautious just at the right time. Only a finely tuned decision-making system can be successful in a risky environment. This is, incidentally, the reason why older test subjects perform worse on similar tests than younger participants. Through the anatomical aging process, the frontal lobe becomes more and more detached from the insular cortex and the nucleus accumbens. As a result, the reward center becomes more heavily influential and older people perform worse in an uncertain market than younger participants who are better able to judge the interaction between risk and chance.[8] Apparently experience isn't everything.

Let me state an interim conclusion at this point: wise decisions are made in the brain through a process of balancing risk and safety. We behave too riskily if this balance is disturbed because our reward center is either too strongly activated or not inhibited enough. From this perspective, risky behavior would seem to be the result of a brain defect (which might be the case, just look at stock market speculation), but risky decision-making is something else entirely. The primary motivation behind a daring decision cannot be attributed to a few loose screws in the brain but, rather, is directly related to the most important driving force for humans.

Half-shocked makes for a bad mood

LET'S SAY YOU are told that you can either, with 100 percent certainty, receive *no* electrical shock by pressing on a particular button, *or* that you will have a 50-50 chance of being shocked by pressing a second button though you will not know ahead of time which it will be. What would you do? That's easy, you might say, no one is idiotic enough to push a potential electric shock button given the choice. And you would be wrong. Humans behave far more erratically than you might think (or have always suspected). In reality, five times more participants reach for the potentially shocking button than for the 100-percent-certain nonshocker.[9] It should be noted that most people kept away from a third, 100-percent-working shocker. This goes to show that we are not totally masochistic.

But why is this? Why do we enter into a calculated risk with possibly unpleasant results if we have another possibility available to us—namely, *not* pressing the 50-50 electric shock button? The most important reason is because the alternative doesn't seem better, at least not to our brain. When you are sitting in front of a potentially hot electrical shocker, your fingers are tingling in anticipation before you've even activated it. You simply want to know if it is working or not. A 50 percent chance of punishment seems better than a 100 percent chance of uncertainty. You thus invest in the tiny bit of electrical shock in order to achieve certainty.

In fact, the stress from uncertainty is measurable. In another experiment, participants played a computer game in which they had to turn over stones in a desert. If a snake was under the stone, they received an electric shock (which just goes to show that neuropsychologists love to use electric shocks in their studies, and their test subjects seem to like it). But the participants did not know beforehand that they would receive an electric shock and, in this case, their stress reaction was greatest (dilated pupils, sweat from anxiety).[10] When they did know with absolute certainty that a snake would soon pop out and they would be given a shock, it was much less stressful for participants.

This goes to show what it is that really drives us up the wall: uncertainty. Anyone who has had to wait for a late train knows what I'm talking about. The fact that the train is late is nothing new. But the fact that you don't know *how* late it is going to be or *why* it is late can really get on your nerves. And, dear train service, your abstract alibis, such as "operation delay," don't really help. That's as good as saying, "The train is delayed because it is delayed." But if we know that the train is delayed by twenty-five minutes, our uncertainty is relieved, and we become slightly less stressed. At least, as long as the planned delay is punctual and not extended further.

And this brings another intriguing truth to light: curiosity is by far the most important human drive, outweighing even our fear and desire to avoid loss. If it were the other way around, humankind would still be roaming the bush in Africa instead of spreading out across the entire globe. Of course, curiosity also requires some sacrifices, but we are unable to do otherwise because nothing is stronger than our urge to experience something new. You could easily suppress or replace other drives. Regular mealtimes? Overrated, just ask management consultants. Seven hours of sleep? Unnecessary, just ask investment bankers. Cuddling time as a couple? Gratuitous, just ask the mechanical engineering students. But if we don't constantly feed our brains new information, or if we feel ourselves dangling in an uncertain state of suspense, that's when we really get up in arms!

The world belongs to the brave

OUR BRAIN RELIES on a constant stream of new input, regardless of the cost. Because the reward for satisfying this urge is usually greater than the looming punishment. This was even the case in the electric shock test, in which the only valuable information gained was whether or not the shocker was turned on. For our brain, curiosity is a value in and of itself. Merely the prospect of experiencing something novel stimulates our reward centers as strongly as if we are experiencing it.[11] In other words: eagerly anticipating gifts is as pleasant for our brain as actually receiving the gift itself. That the most beautiful joy is anticipation has now been proven scientifically.

Furthermore, curiosity drives us to take risks, and this behavior is even firmly anchored in our brain structure. In our brain, the parts of our decision-making system that control and check our impulsive actions (especially parts of the anterior cerebral cortex) are smaller if we are a more Faustian, daredevil type of person.[12] The less brain, the more risk, you could say. Whether or not this correlation is true of greedy market speculators has yet to be researched.

Curiosity is a wonderful thing, drawing us into risk and thus to new shores. In Western culture, however, it has a bad rap. The story is that we were forced to flee paradise because we absolutely had to try fruit from the tree of knowledge. In addition, Pandora pried open the box and unleashed disaster on the world, all because we couldn't contain ourselves. But as a neuroscientist, one doesn't necessarily always consider paradise to be the best place on earth. I mean it might be very pleasant sauntering naked through a botanical garden, but who knows... maybe there are advantages to being outside of the garden? After all, happiness is not a permanent condition for the brain. It is always changing. This is fortunate for us because this is what makes it possible for us to change and adapt (something which has always been necessary in a changing environment). I am pretty certain that our brain would not have remained happy for very long in paradise but would have eventually gotten curious about what else is out there. Unchanging luxury sours at some point, as anyone who has tried eating their favorite food every day might attest. After only about five days, you feel you have French fries or bratwurst coming out your ears. Neuropsychology has finally shown us that to feel "better" with 50 percent probability is better than feeling "good" all the time.

Risky decisions have a bad reputation as well. Perhaps this is not unfounded since our brain can't seem to stick to a single scenario, keeps tipping our balanced decision-making system, and doesn't seem to ever be satisfied with how much it's earned in the stock market. But every later success that we have is based on a risky decision that we have made. The premise for this is deeply rooted in our brain, and it has to do with our search for better conditions. Not for the *best* conditions, because that will eventually get boring for us. Rather, what we want is something that is always a bit of an improvement. Becoming happy is always a little bit nicer than being happy. Our brain's goal is not to be happy but to become happy. That's why it is precisely the unhappy people, the not-yet-fortunate, the hungry, the daring, the risky who make decisions—and change the world. Such risk is—for humans—merely the way out of unbearable uncertainty.

We are always smarter in retrospect

SO HOW CAN we make the right decisions? Wrong question! For our brain, there are no "right" or "wrong" criteria when it comes to decisions. We cannot know beforehand how things are going to end up, and we usually don't have any guarantee of reward or punishment. We don't even know how we would define a successful decision. Only in retrospect does a decision seem to be more or less meaningful. And even then, we are not using a rational system to evaluate our decisions but are rather relying on our emotions as much as we did at the start of our decision-making process. We say things like: "I would never decide the same thing if I had to do it over again," or "This decision was a good one for me."

Every decision is basically more or less risky because it is processed in an uncertain environment. This is, fortunately, something for which our human brain is prepared. Our brain is not interested in being rational and factual about weighing all of the pros and cons and then objectively choosing the best option.

Instead, it generates an emotional impulse that it subsequently backs up with supporting facts. This method of decision-making seems particularly ridiculous in our digitally calculated world. How can we expect to justify a completely emotional impulse and then explain it to others? If you ask people who have just made an intuitive decision for their reasons, they invent all kinds of arguments to justify it. But these arguments can easily be pulled apart, destroying the original (intuitive) decision. People would love to be able to make decisions based on objective facts and criteria, but if they were actually to make decisions based purely on numbers, probabilities, and facts, they would only be able to decide in the manner that algorithms do: predictably and tediously. At the same time, they would put the brain's greatest strength at stake—the ability to make decisions in the face of unclear facts and a context in which a logical computer system would be overwhelmed. In the end, our brain is not made for precision but for uncertainty. Or, to return one final time (in this book) to Goethe: "Better that your decision is nearly right than that it is completely wrong."

Even if you were able to integrate every possibility and eventuality into your decision, the only thing that you would really be able to control would be the exact moment of the decision. You still wouldn't know with certainty what would follow and whether the decision was the right one or not—this is not part and parcel of decision-making.

Far more important than making the optimal decision is being able to take responsibility for whatever decision you have made. This is most likely to happen when all of the brain regions involved in decision-making are in balance. Whenever you throw in an imbalance, it shifts the quality of the decision. You might be too greedy and let the nucleus accumbens take the upper hand. Or you might gather too many facts in a list of pros and cons and thus lean too much weight onto your rational brain regions, in which case you may make the most objectively reasonable decision, but you won't get much joy out of it.

Fortunately for us, we are more than mere biological machines whose only job is to calculate figures and make decisions from a selection of objective options. Doing so is not very easy for us. In the next chapter, I will tell you why, and I will also explain how we can nonetheless land on the best possible selection.

10

SELECTION

Why Selecting Is So Agonizing—and How
We Still Manage to Choose the Best Option

THE PREVIOUS CHAPTER showed us what the brain is good at—namely, pushing itself to reach a decision in the midst of an uncertain context. This is best achieved by balancing our emotions with rational thought. We are best able to make a decision when we have a positive feeling and, at the same time, can pay attention to the facts at hand. But this is not always easy. Especially when it comes to doing something much simpler than deciding: selecting.

I like to eat muesli. For one simple reason: because it is quick (every second that I can stay in bed carries twice as much weight for me), and yet it is still healthy and full of a whole bunch of different ingredients. When I lived in the U.S. for a period, back before bulk sections offered any kind of German muesli, I had to forego my morning muesli routine and opt instead for cereal, which seemed to offer more air than nutrition in my opinion.

Back in Germany at my local German supermarket, I wanted to select the absolute best muesli. This was no easy task since the supermarket muesli aisle has (and I have personally counted) about 118 different varieties on offer. Add to this that my local supermarket is not a very big one, and I didn't even include the eighty-seven different varieties of cornflakes in the total. If you add the twenty-four varieties of milk into the mix, you get a total of 2,832 different possibilities for combining a muesli-rich breakfast! That just goes to show you how quick and easy it can be…

How can we best select something from so many different options? The breakfast muesli question is easy compared to the rest of the 65,000 products that the supermarket carries. When I went to buy myself a pair of pants, the store had 124 different styles to offer me. Together with 169 styles of shirts, I had the possibility of 20,956 individual outfit combinations to choose from, which would have supplied me with a different outfit every day for fifty-seven years. As important as our break-fast and outfit choices are, they are relatively unimportant when compared to some of the more weighty questions in life, such as those we already touched upon in the last chapter: Should I pursue a university education or vocational training program? Start a family or stay single? But even in these significant cases we are met with a seemingly endless array of options. In Ger-many alone (and this would certainly be an even larger number in North America), I could have selected from among 18,044 different university degree programs[1] and could have chosen from among over twenty-one million women to be my partner (theoretically, at least). Who can possibly maintain an overview over such an expansive list of choices?

In fact, you are pretty good at doing this. Every year, over 300,000 new books are published in the United States, and yet here you are reading this particular book and not any of the other 299,999 available titles. Good choice. But how did you do it? The greater the number of choices, the more difficult it is to select from them. If you have options, you have agony, and that is why some people find it exceptionally difficult to make a selection when choices abound.

This is even the case when it comes to political options. During the last federal election in Germany, citizens had thirty-four different parties from which to choose. I personally think this is much better than only having two parties or, even worse, not even having the right to vote. The more options available, the more likely you are to find something that suits your personal preferences and feels like it represents your values. But if the array of choices is too large, we start to experience something known in scientific circles as "choice overload." This occurs when we are pummeled from all sides by too many choices so that we lose our desire to select anything at all, are dissatisfied after we have finally made a selection, and end up regretting the thing we have chosen. Afterward, you think to yourself: I chose poorly. We have all felt this at one time or another, maybe even after a national election.

Our brain seems to get overwhelmed when we push it to make too many choices. Why isn't it more robust when facing a large selection? How can we improve our behavior in choosing? Should we write pros and cons lists? Trust our intuition more? Toss a coin? Let's start by casting our gaze onto the decision-making regions of the brain. Because when we have so many options to choose from, the only option for the brain is to get overwhelmed. In selecting something, the brain relies on its decision-making regions, which are actually much better at other tasks.

Our strength in deciding...

OUR BRAIN IS not hardwired to select the right choice from a vast array of options. Well, strictly speaking it isn't "made" for anything. It is an organ that is always adapting to its individual living conditions. But this is precisely the point. Through this adaptation process, it relies on its decision-making system, which is extremely good at doing one thing: grounding itself in the midst of an uncertain environment—and *not* at comparing numerous options at computer speed.

The previous chapter already looked at the brain regions involved in the decision-making process. The emotion centers of the midbrain—the nucleus accumbens and the insular cortex—are responsible for generating the initial emotional impulse. Conscious-thought centers of the frontal lobe further build on this "gut feeling" by adding a justifiable structure of knowledge and facts. As these regions interact with one another, an active state begins to stabilize, sending a command to move to the motor regions of the brain. Incidentally, this topic—the diffuse back and forth within the brain leading to a final decision and pattern of action—is still being studied by scientists. It is presumed that each decision is translated into a ready-made pattern of action, then is transported as expeditiously as possible beyond a particular threshold in order to maintain its stability. It's a case of survival of the quickest, since the decisions that are successful compared with all the alternatives are the ones that are able to first rally together (or synchronize) enough active brain regions.[2]

The brain is quite well equipped to deal with this kind of decision-making process, using its decisive strengths in the midst of unclear information. Which career path should I follow? How do I want to live? These are such diffuse and uncertain spheres of activity, each having countless variables and possibilities. There's no algorithm in the world that could gather all of these variables under one hat. And yet the brain can "unreasonably" arrive at just such a conclusion by first putting the issue into a rough emotional frame and then continuously comparing it back and forth with facts and experiences, until it works itself step by step to a decision.

...and our weakness in selecting

HOWEVER, AS EFFECTIVE as our decision-making system is, its greatest strength is also its weakness. It is able to wring out a decision by combining emotions and facts. But in order to do this, it must pay a price. The heavier the fact sheet, the more difficult it is for us to make a selection because our decision-making system gets bogged down. Our conscious-thinking frontal lobe has limited resources, after all, and if it gets overloaded with too much input, our balanced back-and-forth system between emotions and rational thought stops working, or at least does not work as well. The system is thrown out of whack. The brain is supposed to be able to compare all of the facts with each other in order to maintain a broad overview. But comparing facts is not at all fun and, in addition, the brain is pretty lousy at this task.

It is usually much easier for us to decide something than to make a selection or choice. For example, a lot of people are perfectly clear in their minds about their desire to have a family someday, with a loving partner and one or two kids. The decision has been made, and it's a big one! But then things get a whole lot harder. Because the next step is choosing "the right one" from several hundred thousand potential partners. This can get overwhelming, to say the least.

In such cases, computers have the upper hand on us. Comparing and contrasting a couple of potential life partners is no great art if you really think about it. All you need is a lot of information and good math skills. If you tell a computer to rate certain factors such as appearance, hobbies, interests, and personal values, even a soulless algorithm will be able to select your potential mate. This is true even if you have 100,000 candidates instead of only 100. Sound too impersonal and callous to you? Maybe you think an algorithm can't possibly know the most important inner traits of your potential partner? Perhaps, and yet millions of people use the dating app Tinder, placing their romantic fates into the hands of an algorithm. And they do so in spite of the fact that nearly half of all Tinder users are already in a relationship.[3]

Remember: making a selection is not the same thing as making a decision. A matchmaking algorithm is able to filter out several options (potential partners) from a long list of alternatives. But it runs into trouble when deciding which partner is the right one. Conversely, the brain gets overwhelmed trying to select from too many options. But it has already made the decision that it wants to live life with a partner at one's side.

So, you see, we are very good at making a decision, but this is also the reason why we have trouble choosing from among many similar options. This weakness—selecting the right option from a large range of choices—is known in scientific research as "choice overload." And researchers have figured out exactly which conditions make choosing so agonizing.

Our Achilles heel of choice

I AM NOT the only one who has a hard time choosing the right muesli from the grocery store. Participants in a psychological experiment from 2000 faced similar difficulties selecting their favorite jam. Researchers set up two different scenarios in a supermarket: one display table featuring six different fruit jams and another with twenty-four varieties. Over the course of several days, they observed how random shoppers reacted to the jams on display. The results should not come as a surprise: shoppers remained standing in front of the display table with lots of jams for a longer period of time but overall bought less. Of course, the display with abundant jam varieties was more attractive to the eye and lured passersby over, but in the end shoppers were simply too overwhelmed by the selection. Conversely, of those passing the display with only six jams, 30 percent purchased a jar, versus only 3 percent purchasing a jar from the table with twenty-four varieties.[4]

The jam experiment was very inspirational, both for science and the market. At the start of the 2000s, Procter & Gamble reduced their selection of hair shampoos from twenty-six to fifteen varieties and saw a 10 percent increase in turnover.[5] Take away the burden of comparing alternative items and sales go up. This is a concept that discount grocery stores have already perfected. While a normal supermarket can easily stock 100,000 different products on its shelves, the discount store Aldi only offers around 1,300. Of these, there are only three varieties of muesli, instead of the average 118 in German supermarkets.

As logical and plausible as it seems that we would be overwhelmed by too many options, it is actually quite difficult to replicate the phenomenon of choice overload. Eight years after the first study, a Swiss research group replicated the jam experiment—and failed to find the same results. Subsequent reviews showed that choice overload is not as clear as one might think.[6] It turns out we are not always overwhelmed by a large selection, rather it depends on the surroundings in which we must make our choice.

Too much chocolate does not happiness bring

RESEARCHERS LOVE TO study the ways in which chocolate can affect our behavior because it's easy to find willing participants. When looking at how participants reacted when told to select one kind out of a large variety of different chocolates, there was an indication that a large number of options had the effect of decreasing the selector's pleasure. Researchers divided the participants into three groups. The first group was told to choose the most delicious chocolate from a selection of six different varieties. The second group was given the same task with thirty different varieties. The third group had bad luck and was only given a single variety to "choose."

Once again, researchers found that having a large selection is not necessarily equated with happiness (even though the participants got to eat a lot more chocolate). Participants in the second group who had to choose the most delicious chocolate from thirty varieties (and were then allowed to take home their choice) were ultimately less satisfied with their selection than those who'd had only six varieties from which to choose.[7] In addition, it was much harder for participants from the second group to decide which sort was the most delicious. This is no great surprise—after all, how could one possibly choose from such a large selection? You would have to try every single kind! But by the time you've reached chocolate variety number 22, it will be hard not to get sick of the different cocoa blends. Chocolate number 22 is a tough act to follow. And after having sampled chocolate number 19, who can remember chocolate number 12? It's not surprising that only one out of every eight participants in this group decided to take home any chocolate as a reward for their participation. Instead, they chose to take home the value of the chocolate bar in cash. Of course, the unhappiest participants were those in the third group, who had only been allowed to "choose" from a menu of one chocolate variety. Of these participants, the same number opted to take home the chocolate as those in the group with thirty varieties. Which goes to show that too few options is just as bad as too many. We are able to appreciate having some options, but at some point, we get overwhelmed when we have too many alternatives. This doesn't only apply to chocolate, but also to electronic gadgets[8] and market shares.[9] When the number of choices gets too big, people simply stop deciding.

This changes, however, if one is familiar with or at least knows something about chocolate (or market shares). In one experiment, participants were asked to indicate which type of chocolate they preferred before they then had to select their favorite variety from a small or large array of options, respectively. This study showed that the more one knew about what they wanted, the easier it was to choose their favorite from a large selection, as well as to be happy about their choice in the end.[10] An abundant selection is thus not necessarily a bad thing—it is only a problem if you don't have any idea about what you want. If, on the other hand, you have a clear objective, it's a whole lot easier for your taste buds to compare thirty varieties of chocolate to your favorite flavor.

Selecting for others

A LARGE SELECTION is not always a bad thing. For example, when you aren't deciding for yourself but for someone else. Let's say a colleague asks you to choose a snack for them at a big snack machine. Let's even say that you know your colleague pretty well (which is important in this case). Would you be more inclined to go for a snack machine with a large selection (thirty-six different types of snack) or the one with a smaller selection (only six different snacks)?

If one has to later justify one's choice of snack machine to someone else, most participants will choose the snack machine with more options.[11] We prefer to be able to argue that we selected from the larger range of options. On the other hand, if we have to justify the particular snack that we have selected, we prefer to choose from the sparsely stocked machine. At least this way, if the granola bar gets lodged in our colleague's throat, we can argue that we didn't have that many choices.

The more we have to justify our particular choices, the less we prefer a large selection. Because a large selection forces us to explain why we chose *this* particular option from all of the available options and, on top of that, we have to do all the extra work of reviewing and comparing the different options. This dynamic is especially tricky when we are socially observed during our selection—for example, when making a donation. In a study from 2009, researchers gave participants the opportunity to choose a single aid organization from a selection of five, forty, or eighty different organizations to which to donate one Euro or more (from their own pockets). One group was asked to justify their choice while the other group was spared this procedure. Interestingly, the pressure of justifying their choice led participants in the first group to donate less money if they were faced with choosing from a selection of forty or eighty charities.[12] It turns out that this requires one to argue for donating to help orphaned children over those with leukemia. There's no joy in a moral dilemma.

In principle, having a large selection is not a bad thing. It can be quite the opposite if we know exactly what we want or are familiar with the field in question. In this case, a wide array of alternatives can be rewarding. Shopping is fun because we are better able to realize our specific goals by selecting from among a plethora of options. If a car-crazy motorhead is strolling through a used-car lot, the inventory is never large enough. They have already long ago done all the preparatory work of gathering information, comparing, getting familiar with all the relevant facts, etc. (and, in any case, a search machine has already filtered out the options). My mother, on the other hand, has only a single criterion for a car: "that it drives." Trying putting this somewhat imprecise selection criterion into an online database filter for used cars. Good luck!

Cost pressure is also in the brain

QUICKLY COMPARING DIFFERENT options is such an arduous task that we are happy if we only have to face a small selection. Remember that the decision-making system rationally analyzes multiple facts and impressions in the frontal lobe in order to justify our emotional impulses; however, it has a limited capacity. When this system is put under pressure because it cannot quickly process all of the input, we start to experience what neuropsychologists call "cognitive dissonance." In other words, our expectations and reality clash, and we are left unsatisfied with our selection when forced to make a selection under time pressure.[13]

If you have to decide for something, you also have to decide *against* something else. Thus, the greater the selection on offer, the more options we must actively choose to reject. A process of elimination in which we have to justify every one of our non-choices is a difficult task. Let's say you are sitting across from the host of *Who Wants to Be a Millionaire?* and that you are no longer faced with four but rather twenty possible answers. The 50-50 option (which eliminates half of the incorrect answers) will not help you in this case because the game would still require you to know and to be able to eliminate the nine other false answers.

Add to this the fact that reality consists not only of right and wrong answers but also of half-right and more-right-than-wrong answers. The more of such options that we face, the more we tend to regret our decision. The reason for this is a phenomenon which scientists call "alternative costs," also known as opportunity cost.

Anyone who has ever watched a casting show is familiar with alternative costs and the reasons why the decision becomes harder and harder the longer the season goes. At the start of the season, Tyra Banks doesn't have a problem kicking out the most incompetent models. But when only five models remain, things start to get complicated. "This was a really tough decision to make," the judges say. "We didn't come to this decision easily." The reason for this is clear: in selecting a single top model, you have to make a choice *against* all of the others. In addition, the top-model candidates start to look more and more alike as the series progresses. They all look attractive, have mastered the catwalk, and know how to pose for the camera. The more their abilities resemble one another, the harder it is to distinguish them based on the selection criteria—not unlike the twenty-four types of jam that were different varieties though they were all somehow alike. Thus, when Tyra Banks chooses the next top model, she is taking a risk: Who is to say that the other candidates wouldn't also go on to have equally amazing careers in modeling?

It certainly puts a damper on the overall mood, because even though we think consciously and rationally about such alternative costs, they nonetheless affect our emotional brain centers. Recall that our rationally thinking frontal lobe is connected to our emotional limbic system and can steer the activity of our pleasure center (the nucleus accumbens). That's why we get so nervous about having a large selection, because even before we've made a choice we worry that we might regret whatever we are about to select.

A further issue is that our decision-making system learns from the past and is thus always evaluating what it can do better in the future. If you have made a choice with which you are dissatisfied, it's your own fault. Because of this, we tend to prefer choosing from a smaller selection if we are unfamiliar with the territory. But if we are selecting something for someone else, it's a different story: we expect our doctors, lawyers, and investment or tax advisors to provide us with the full array of options because then they are the ones who have to deal with the alternative costs, not us.

Avoiding the agony of choice

SO, WHAT CAN be done to improve our competency in decision-making and selection? We can hardly expect that we will have fewer options to choose from in the future. Not to act is also not an option. And anyway, making a selection can be fun as long as you don't let yourself get too overwhelmed.

Trick #1: Narrow your goal.

Research studies are constantly finding that it makes an enormous difference whether we have a concrete picture of our goals or only a vague idea what the outcome should be. When the study participants in the chocolate group with the large variety of options already knew their favorite type of chocolate, it was much easier for them to compare the other chocolates. The result is that we are much less likely to be overwhelmed because we don't have to compare every single option with every other option—but only with our fixed goal. However, watch out that you don't fall for any oversimplification tricks. Salespeople are very familiar with this method and use it to add an additional, more dominant offer to the mix in their effort to help you decide. Their product could be a particularly bad or a wonderfully good product. You will tend to compare the rest of the options with this product and be easily drawn along to your purchase.

For example, if a florist wants to sell more bouquets, they put out some long-stemmed red roses next to the other flowers. It doesn't matter if the customer likes the roses or not, such an eye-catching display enhances a customer's purchasing behavior, and they are more likely to choose any option.[14]

Trick #2: Be satisfied.

Those who are most sensitive to selection overload are called "maximizers" in psychology lingo. These are the people who do everything in their power to make the best possible choice. They analyze, compare, brood over the problem, ask experts and continue their search, sparing no expense or effort to make the absolutely best selection—and then end up unhappy with their choice. This can happen, for example, when you are job hunting. Researchers who studied recent university graduates who were searching for work uncovered the following pattern: those who indicated in a questionnaire that they were looking for the absolutely best possible job did, in fact, find positions with a 20 percent higher average income. However, they were less happy than those who had searched for a job in a way that was quick and uncomplicated.[15] One possible explanation could be that over the course of one's perfect decision-making process, one is also faced with the many different possible options. Maximizers may have a higher income at the end of the day, but they are also aware of the alternatives they had to reject. While the maximizer believes they could always have potentially selected something better, the satisficer (the psychological term for the satisfied person) hardly gives it a second thought. What he doesn't know does not hurt him. What this means for us practically is that we should take care to make up our minds and decide what's important before we start trying to make a specific selection. What is more important to you: money or happiness? If you wish to be happy, stop wasting so much time searching for the perfect solution and don't pay too many high "alternative costs."[16] Of course, a bit of thoughtful consideration is always a good thing, but don't pass the point where making a choice turns into overthinking.

Trick #3: Decide important things with your gut.

The more options we have to compare to one another, the longer the decision-making process takes, the more we are aware of alternative costs, and the more dissatisfied we end up being with our final decision. Try the following trick to overcome your choosing challenges. You might recall from the beginning of the chapter that every decision more or less starts with an emotional evaluation: What is the maximum reward and what do I have to do to get it? In principle, your brain looks for facts to back up your emotional goal. So why not just save yourself the justification and make a quick, emotion-based choice? This works especially well if the array of choices is vast, and the decision is a long-term one. If you are buying a towel and faced with a broad selection, there's really not much that can go wrong with a towel, so choosing should not be a problem. But if you are in the market for a house or a car, that's when you really start to ruminate. In these cases, however, it is much more important to be able to justify your decision emotionally rather than rationally. Facts will help you to justify your decision to others, but only you will be able to justify it to yourself if you have an intuitive gut feeling. This is why people tend to be happier and to make better decisions when they don't think too consciously about buying a car but instead let their intuition take the reins. One specific experiment asked one group of participants to select a car to purchase based on four categories. The second group was asked to make the same selection based on twelve different criteria categories (such as trunk space or mileage). The more diverse criteria the car purchasers were asked to consider, the better and happier they were at choosing if they were told to solve a word puzzle in between the car presentation and their final decision.[17] Thus, the next time you are at a car dealership, complete a crossword puzzle before you start ruminating. Or set yourself a deadline, defer your decision by a few days, and distract yourself with some leisure time. Intuition is not always irrational; it can be a much better alternative to conscious rationalization in the midst of a complex decision.

Trick #4: Resist the excess.

You have read that the main problem of decision-making lies in the necessity to compare many options. And the more diverse these options are, the more energy it takes for us to compare them. At some point we get overwhelmed and simply throw up our hands. You can see this phenomenon when single people speed date. The more diverse the selection of potential life partners, the less chance that a single person will want to see any of them again.[18] You can remedy this dilemma by creating categories. This strategy helps you not to have to make a separate decision for every single option. It also helps you to be happier in your choice. When study participants were asked to choose a single magazine from a selection of 144 magazines, those who were first able to choose from fourteen presorted categories were much happier with their final decision than those who could only choose from three categories ("Women," "Men," "Various").[19] What kind of categories you come up with is not that important. The main thing is that you give yourself a few (which shouldn't be too difficult for you, since the next chapter shows how easy it is for us to construct thought patterns).

Trick #5: Up the pressure.

It is not so hard to make the best possible choice from a broad selection if we have organized, and therefore simplified, the possibilities according to our personal criteria. It's not all that important for the brain whether the final selection is objectively the best one. If this were the goal, our brain would have to analyze and compare far too many individual parameters, which it does not have the capacity to do. It is much more important that you are able to live with your decision, and in order to do this, you have to know how you feel about it. If you have ruminated for a long time over a choice and thought through all of the alternatives, your brain has already long since made up its mind. The more familiar you are with an issue, the more important it is to recognize your emotional and intuitive core.[20] This core is often concealed beneath a heap of facts and rational arguments. In this case, it can be helpful to put yourself under a bit of pressure. Toss a coin, pull a card, roll the dice. In the moment when you release the dice, you already know which number you want it to land on, or whether it has landed on the wrong number.

Of course, I have given up my habit of rolling the dice each time I go to find a new type of chocolate muesli in my local supermarket. Instead, I remind myself about my biochemistry studies and have thus created a valuable muesli selection criterion for myself: the protein-to-fat ratio should be around at least 0.9. If it is, I toss it in my cart. You can decide for yourself whether or not this is a good thing. But for me, the stuff has to taste good as well. Because when it comes to the brain, feelings are everything.

11

PIGEONHOLING

*How Prejudices Can Help Us, How They Can Harm
Us—and How to Avoid Stereotypical Pitfalls*

*Jingles that rhyme stay in brains
Helping products to make gains
A punchy slogan seems to cling
It's the power of patterning
But jingly rhymes trick the head
Making us settle for the simple thread
In this chapter you'll see defined
The pros and perils of the biased mind*

If you see some information, cluster it in aggregation!

NOT ONLY DO we tend to perceive rhyming advertisements as more honest and more important than nonrhyming ads (even when the content is the same),[1] we also employ other subconscious cues to mentally pigeonhole or build stereotypes. And we often do this too rashly:

The young man named Wyatt grew up on a ranch at the edge of the plains. At a young age, he joined the local line dancing club. His rifle collection is legendary, and he plays the slide guitar. His favorite food is barbequed ribs, and his favorite football team is the Dallas Cowboys. Last summer he moved to Austin to pursue his career.

Which do you think is more obvious: That Wyatt is an aspiring car salesman? Or that Wyatt is an aspiring car salesman who has already been to the rodeo this year?

Of course, as an educated and worldly reader, you won't be easily fooled. The chance that Wyatt wants to be a car salesman is much greater than that he has also been to the rodeo. And yet most people tend to believe that the second scenario is much more plausible because it simply fulfills more of the aforementioned (and unmentioned) categories. Did you notice that some of the details of the story were *not* disclosed but that you nonetheless formed your own picture of them? The story didn't say that Wyatt grew up in Texas, only that he moved to Austin. But you probably subconsciously decided that this was the case because it fits the stereotype of a typical life story of a Texan (okay, and "Wyatt" is not exactly an East Coast name).

The psychologists Amos Tversky and Daniel Kahneman conducted a similar experiment in 1983[2] that provided a crucial insight to a particular weakness in the brain, which they called the "conjunction fallacy." This very effective psychological effect pulls the wool over our eyes more than we would like to admit. We are in the habit of permanently bending things to fit into our world view.

Did the ball fall on red five times in a row during roulette? Then it's time to put your chips on black because every series has to be interrupted at some point. Did the share prices drop for two weeks straight? Then surely the trend is going to reverse course now. A hurricane named "Katrina" is forming off the coast? It can't be that bad—after all, it's not named Kevin, the male name that was voted the least popular name in Germany, along with the female name Chantal.[3] If you were in Germany and happened to see a Kevin walking down the street holding hands with his platinum-blonde girlfriend in stilettos, you might not automatically think both of them are bound for Harvard.

We draw connections, come up with correlations and narratives, pigeonhole others and create stereotypes that are often not true. In short: we make decisions too hastily and rashly, and particularly without taking the time to think through our opinions and to perhaps prove them wrong. It seems much more important to be able to form a thought pattern that can consistently explain the world to us. And if the world doesn't fit into the pattern, we make it fit by acknowledging only certain factors of our choosing.

Such thought routines offer the enormous advantage of speeding up our thought processes. This is great when it comes to most daily situations because we don't have to waste time or energy on surplus consideration. Instead, our brain can simplify and continue to refine its mental categories even further. Stereotypes help us to think ahead, not to think about. In this way, we are able to reach a judgment very quickly. But it's important to be aware that these are prejudices because we are relying on our thought patterns and experiences and not on objective criteria. Such thinking can lead to serious wrongdoing. If we are lucky, we might only stumble into committing a slight faux pas because we didn't pay attention to proper etiquette. But worst-case scenarios in the stock market or with regard to racism or resentment often have as their base a twisted pattern and stereotyped way of thinking. Pigeonholing might make our lives· easier, but it can also close us in—into an intellectual prison, a mental box. How can we get out again?

Mental taunting

I SHOULD EXPLAIN here that our brain is really good at constructing stereotypes. To be more precise, this is the primary activity of our brain. The study by Kahneman and colleagues from over thirty years ago has already shown just how potent and seductive pattern thinking can be. Interestingly, in the early 1980s, researchers discovered an effect in the brain that explained very elegantly what happens when we form mental boxes.

Imagine that a stand-up comedian is performing in Boston. He comes onto the stage, cracks a few cheap jokes about Hollywood celebrities—Justin Bieber or some other easy target. Punch line, laughter, short applause, next joke. In other words: the audience responds according to a particular social pattern by giving the comedian a brief positive answer, the applause. But now the comedian suddenly puts on a Yankees baseball cap, tears off his jacket to reveal that he is wearing a navy blue and white striped jersey, and starts to sing the Yankees theme song. He probably won't get much further than that because he will be booed and heckled off the stage by the Boston Red Sox fans. Obviously, the Boston audience's expectations were not met and the result is resoundingly clear. They respond with annoyance and negative reactions. If there is something that does not fit in very well (such as a Yankees fan in Boston), it is going to be refuted initially.

Something similar happens in our brain. It is permanently building a frame for expectations and then checking to see whether or not everything fits with these expectations. Our neurons' reaction is not unlike the way that comedy-club goers will sit in front of the stage and clap or boo at a comedian. We don't have to be aware of what every individual cell is doing because the overall result is much more important. Together, all of the neurons create an electrical field that is strong enough to measure. By positioning electrodes on a subject's head, researchers are able to observe minute, millisecond fluctuations in these electrical fields that show just how strongly the neurons have rallied to applaud or boo collectively. This procedure is called EEG, electroencephalography, or "electrical brain writing."

The most well-known boo signal in the brain is the N400 signal.[4] Because it produces a negative electric field deflection, the letter N is used. The number 400 is designated because the signal occurs about 400 milliseconds following the key stimulus. If there is some component that does not fit into one's frame of expectations, the neurons respond with the N400 signal.

For example, if you have to complete a list. In studies on this topic, participants are often presented with the following words with one second between each while their neural electrical field is measured:

twinkle

twinkle

little

pudding

Yes, pudding doesn't fit. "Star" would have been a much better fit to complete the well-known children's lullaby. Not surprisingly, the N400 signal shows up measurably in participants' brains following the word "pudding." When the word "star" was shown instead, participants' neurons remained relatively calm.

The benefit of pattern thinking

OUR BRAIN IS constantly forming little boxes into which we try to mentally stuff everything that crosses our paths. If something doesn't fit, our neurons protest, booing like the Yankees fans at the Red Sox. Since most things in our lives tend to follow a routine, this is usually a practical mode of being, since it allows us to quickly and flawlessly navigate through a familiar landscape. We simply unroll the course of action that fits to the specific situation and we're off (for example, laughter and applause in response to a joke on the stage).

If, however, we suddenly arrive at a conflict and the matter at hand does not fit into our frame of expectations, we immediately become alarmed. In this case, we have two different options: we can either change our stereotypes and adapt them to the new conditions (wait and see whether the Yankees comedian might have a good punch line after all). Or we can turn away from the offensive stimulus (heckle the guy off the stage and swear never to attend one of his shows again). Unfortunately, humans tend to choose the latter, much-less open way of being—and that leaves us more vulnerable to deception and a skewed world view. There is nothing more important for our brain than to maintain the stability of its thought stereotypes and patterns. Every one of us keeps patterns like this on hand for all different kinds of situations. We have a pattern for driving a car, a pattern for eating breakfast, a bedtime pattern. Predetermined patterns simplify our lives. This is a good way to get along in life when situations are clear and repetitive.

Our lives are an efficient series of these kinds of patterns. Simply by applying them to the corresponding situation, we are able to avoid making mistakes. This way of thinking is absolutely fundamental to the brain. As soon as we arrive in the world (and even before), our brain tries to recognize patterns in its surroundings and thereby to set up stereotypes and categories. We can't do without them, and yet this quick and tidy way of thinking comes with two disadvantages. Firstly, we construct our mental boxes too hastily by drawing correlations where none exist. This is dangerous because it leaves us vulnerable to seduction and deception (we will look at this shortly). Secondly, we are too slow to deconstruct our mental boxes. This is dangerous, too, because this leads to prejudice and resentment.

Bringing together what doesn't necessarily belong together

OUR BRAIN IS permanently trying to recognize correlations and to mold them into narratives. To this end, it references every imaginable stimulus, every teensy little detail, to be able to pigeonhole it into a workable thought pattern. What's bad about this is that the subtler these details are, the easier we allow ourselves to be fooled by false correlations and false pigeonholes. Let me give you some examples.

Would you like to convey your message in a way that is both catchy and convincing? Then be sure to write clearly! If people read a cookbook written in an easy-to-read font, they also rate the actual cooking process of the recipe as quick and easy to follow.[5]

But if the text is written in a hard-to-read font, study participants think that the cooking process must also be tedious and complicated. Illegible texts are generally less convincing than clearly written lines. It's bad luck if you don't have very good handwriting. A reader will inevitably rate the contents of scribbled letters as worse than a cleanly written text. I am extremely grateful that I do not have to turn in a handwritten copy of this book. The reader's comprehension of this book would hardly be able to withstand the effects of my awful handwriting.

Would you like to come across as warm and caring to the person across from you? Put a cup of something warm into their hand. Because when people hold a cup of warm coffee in their hands, they consider the person across from them to be warm and friendly.[6] Whether or not we find the person across from us to be "hot" when we are holding a hot drink in our hands is a subject that has not yet been studied. Nor has the opposite, whether we will seem particularly "cool" if the other person is holding an ice-cream cone.

Do you see a spark of truth in the following phrase?

"Think like a man of action, act like a man of thought."

Really? If this is the case, you are most likely in a good mood and you trust your intuition. When people are asked to report how much meaning they assign to vague aphorisms such as the one above, it depends largely on their mood. The study concluded that the better mood a person is in (and this was a very important factor), the more they believe they are trusting in their intuition when recognizing a deeper meaning in ambiguous sayings (even when there isn't any). Those who are in a worse mood do not see any meaning in such aphorisms.[7] Therefore, if you don't see any sense in the above saying, you must be a grouch. Or maybe the saying is simply completely meaningless? It's probably the latter...

In the same study, researchers asked the fans of a U.S. football team following a crucial defeat of the season whether it held any deeper meaning for them. Still reeling from the previous day's awful game, the most depressed fans saw less meaning in it than those who were not quite so demoralized. For the saddest fans, the defeat was simply something that happened, about which no one can do anything. Sometimes we lose, sometimes the opposition loses, they said. It's quite a practical solution, if you think about it. If everything goes well, there's a meaning in it. If it doesn't, it's senseless. This is really the only way it's possible for me to root for the Oakland Raiders year after year.

This is especially true in the experiment if the participant is an intuitive type of person—that is, someone who tends to put a lot of weight in their "gut feeling" before deciding. Of course, neuroscientists know that this apparent "gut" feeling actually takes place in the head. Although... I have to correct myself. Because when researchers conducted a study on judges' rulings, they realized that the robed magistrates granted prisoners day parole release almost zero percent of the time if the case was brought to their court just before lunchtime. After the judges ate their good meals, nearly two thirds of prisoner requests for day parole were granted.[8]

The prejudices that we form and the correlations that we see thus have to do in many cases with small subconscious details. When people are shown pictures in which an object (such as a ship or a house) is hidden, and then are shown pictures consisting of nothing but a chaotic jumble of lines, they suddenly start to see objects in the chaotic pictures even where there are none. This is only true, however, when researchers have first generated a sense of lost control among the participants by instructing them to recall situations in their lives in which they have lost control.[9] The same thing happens to skydivers. Right before they jump out of the airplane, skydivers report seeing objects in photos that consist of nothing but chaotic lines. But if you ask them when they are not under pressure or stress, they are much less likely to fall for such misperceptions.[10] In other words: the less control we feel we have, the more correlations we see, even where none exist. No wonder there's a rise in conspiracy theorists in times when the economy becomes more uncertain.[11]

Our brain spends all its time seeking out correlations, which can then be tied to narratives. We, in fact, value this so much in our society that people who quickly draw connections are seen as especially intelligent. No intelligence test is complete without a logic puzzle. See if you can complete the following sequence of numbers:

$$1 - 4 - 2 - 5 - 3 - 6 - 4 - ?$$

Someone quickly able to figure out that the answer is 7 is considered very clever because this shows a nimble ability to notice the basic correlation between the numbers. But do you see any logic in the following sequence?

OXXXOXXXOXXOOOXOOXXOO

If so, chances are you're a basketball fan. Admirers of LeBron James and Stephen Curry often believe that shots and misses occur in small clusters (a player sometimes has a run). So it's possible for someone to see a pattern in a completely random sequence like the one above, even if there is no pattern.[12]

We get so fixated on fitting things into mental boxes that we often fail to realize not every correlation is backed up by logical reasoning. We overinterpret our world and allow ourselves to be led astray.

Intolerance is a hindrance

WE ARE PRONE to forming biases too rashly. On top of that, we are too slow at deconstructing them. We hold on tightly to our world view. And though our simplifications of thought can sometimes be good, they turn dangerous if we cannot let go of them.

Hannah, a fourth-grader, comes from a middle-class family. In a video, you can see how clean her room is, that her parents both have solid jobs, and that they live in a posh neighborhood. On a test at school, she is able to solve some difficult problems with flying colors but has difficulty on a few simpler tasks. In your opinion, has Hannah performed better or worse than her classmates on this test?

When this scenario is presented to a group of study participants, they predict Hannah's abilities to be higher than average. This in contrast to a second group, whose participants have been shown the exact same test results but who have previously learned that the girl comes from a down-and-out lower-class family. In this case, participants rate her abilities as weaker than the rest of her class. Significantly, Hannah's test results were the same, though ambiguous, in both scenarios.[13]

It's sad, but true. The more ambiguous the situation, the more tightly we cling to our mental boxes. Even to the detriment of Hannah, who is assessed differently in spite of her consistent test performance. The problem lies with the fact that we do everything in our power to maintain our current world view. We aren't on the lookout for the flaws in our thinking, but for confirmation. The ultimate result is that we get so lost in our own biases that we are no longer able to judge freely. Instead, we build ourselves into a mental prison that we go right on constructing. This isn't anything new. We have always thought in stereotypes. But in our current digital world, we seem to attach even more significance to affirmations of our own world view than ever before.

Social opinion ghettos

WE KNOW WE form hypotheses about the world and then form our mental boxes, which we hope to see affirmed. What are we doing? We call into the canyon whatever we want to hear echoed back again. In our digital day and age, we of course don't need a canyon. We have the echo chamber of Facebook, which repeats back precisely what we want to hear.

A study from 2016 showed how this phenomenon played out among Facebook users who were either members of conspiracy theory groups on the one hand, or science enthusiast groups on the other. Interestingly, the results showed that both the conspiracy and the science news items shared a common dynamic:

story headlines that were particularly eye-catching were quickly liked, shared, and commented upon among both groups. However, while interest in the science news stories eventually waned over several hours, the interest in conspiracy theories continued to gain prominence and multiply over time.[14] This leads to echo chambers in which our own perspectives are reaffirmed over and over again. A trick made possible when most of the members of a conspiracy theory group share the same perspective, and Facebook's algorithm places more weight on similar opinions of the group members. Such an algorithm-filter bubble is very problematic because when we see our own opinion mirrored back to us from someone else, it gains more objective weight for us ("someone else said so too") even if it is a subjective opinion.

In addition, we shield ourselves more and more from any possible counteropinions. An earlier study looking at sentiment on Facebook posts from 2015 showed that comments on science and conspiracy news items became progressively negative over time.[15] This deterioration of the mood was especially evident in the conspiracy theory groups. After a thousand user comments, negative and aggressive comments outweighed others. It might be good to note that piecing together conspiracy theories makes for a bad mood. Although it should also be stated that 45 percent of all the Facebook posts that were studied (which was over one million) were negative in tone. So much for social media helping us to feel better!

Never before in history has it been possible for us to surround ourselves with so many similar opinions, world views, and like-minded people. Dating agencies and websites base their entire business concept on this idea. Your "match algorithm" is programmed to bring together people who share similar hobbies and tastes. Better beware of forming relationship and opinion ghettos in which we are shielded from any outside views! Since we are already predisposed to seek confirmation for our biases, putting up these kinds of walls to surround ourselves only with those who share the same views will only make us less open for anything new.

Consumer junkies through clichés

THERE ARE GREAT advantages to the practice of adapting our biases every now and then. Stewing in our own juices for too long can lead to extremely faulty thinking. This becomes apparent when one studies the effects of stereotyping on people's thinking abilities. The more we get stuck in our prejudices and routines, the more difficult it becomes for us to think outside of our familiar thought patterns.

In order to test how much people rely on their biases and thereby fall into thought traps, researchers often use so-called "tests for cognitive reflection." Maybe you already know this brainteaser from puzzle books: A pencil and a piece of paper together cost $1.10. The pencil costs one dollar more than the piece of paper. How much does the pencil cost? The more deeply entrenched we are in our societal stereotypes, the sooner we fall for the trap and reply: "One dollar."

When researchers showed this test on February 14 (Valentine's Day) to U.S. participants, the color of the screen on which the question was written (with a pink or white frame) played a role in their responses: when the frame of the screen was pink, responses were notably worse because the participants were more likely to fall for the obvious, though false, answer. Apparently, the traditional cultural pattern (pink on Valentine's Day) reinforced the effectiveness of the participants' biases. They were no longer able to think freely. When the test was given a week later, the color pink had no influence whatsoever.[16]

The same result can be obtained by showing participants typical wedding photos immediately before asking them to solve the puzzle. If the bridal couple was dressed traditionally (she in white and he in a black suit), participants make mistakes more often than when they were shown a less traditional bridal couple (she wearing green, he in a purple suit). This indicates that even just looking at stereotypes makes us dig in and become inflexible in our way of thinking. And not only this. Researchers also studied participants' consumer behavior. And strangely, those who had looked at images of typical, traditionally dressed bridal couples were much more prepared to buy completely arbitrary products such as a duvet or a shovel. Not that a shovel is useless, by any means. But what kind of a wedding gives us the urge to go out and buy a shovel? The more unusual the pictured wedding ceremony was, the more participants were immune to impulse purchases. Maybe this is why so many wedding gifts at traditional weddings are so silly. If you are planning your nuptials and really want to get some nice, thoughtful gifts, be sure you wear something unusual to the altar. If you say your "I dos" in shorts and flip-flops, you might risk offending your entire guest list, but you might also be spared the set of pots.

Escaping the stereotype trap

THINKING IN STEREOTYPES is foundational for our brain and is furthermore very practical for us to function. Unfortunately, we are so quick at constructing our biases that we draw false correlations. In addition, we do not evaluate our thought patterns often enough but rather become entrenched in our mental monocultures. This leaves us vulnerable to overhasty errors of thought.

What can be done? First of all, we should be clear that we have a weakness for correlations. We are always on the lookout for connections, which we then overinterpret. Just because something occurs at the same time as something else does not mean that both share a common cause. Correlation is not necessarily causality. For example: What is one of the greatest risk factors for contracting Alzheimer's disease? A study from 2010 showed that it is caring for a partner with Alzheimer's. Of the 2,400 study participants, those who cared for their sick husband or wife (and not just a relative) were six times more at risk of getting Alzheimer's.[17] This phenomenon is known as "caregiver burden." Scary, isn't it? The more one cares for a sick partner with Alzheimer's, the more likely one is to get sick themselves. Is there a correlation? A causality? It's hard to say. Alzheimer's is not a contagious disease, this much we know. And yet we tend to too quickly draw causal correlations where they do not necessarily exist. Asking ourselves whether something is simply a chance correlation or whether a real reason exists for something is one of the most effective defense strategies against falling prey to fictitious connections and overhasty biases. In the case of the Alzheimer's study, one possible explanation was that the partners who cared for their sick spouses had lived for many decades with a similar lifestyle and under the same conditions that created increased risks for getting Alzheimer's disease. Or, do you have a better explanation? If you do, could you please also help scientists to explain why the risk for brain tumors increases if a person has at least three years of higher education or has an above-average income?[18]

Be wary of being too quick to judge. The easier it feels to make a judgment, the more likely it is that this judgment is based on a bias or thought pattern. We, unfortunately, only possess an alarm system that wants to keep our biases intact (the N400 signal), and none to indicate how often or how falsely we rely on our biases. This is why we should be particularly careful whenever we are making an ad hoc decision. It might go well enough but sometimes we simply forget to look for counterarguments. Instead, take off your rose-colored glasses every now and then and play the devil's advocate. Try tearing apart your own ideas. This is much less painful than when other people or reality does it for you.

At the end of the day, we too easily allow trifling matters to sway us into forming hardened thought patterns (just remember those hungry judges!). But when unconscious cues urge us to behave and think with bias, we can turn this weakness into a weapon: by exposing yourself to an unfamiliar environment and fresh impressions, you color your thoughts in a whole new way. Add in other aspects such as a full belly, sun instead of rain—and suddenly the same factors start to take on a new light. Don't be taken in by the next subtle deception. Face your biases by creating a whole picture from the sum of your judgments. Or, as Abraham Lincoln said: "I don't like that man. I must get to know him better."

We are not usually immediately sympathetic to unfamiliar opinions and alternate points of view. If someone contradicts us, we don't think it is a good thing. But it is precisely this diversity of ideas that helps us to make better decisions. Someone who is only wrapped up in their own perspective will end up in a bind, leading to poor decision-making. The effectiveness of deliberately changing one's views was highlighted by a study in 2012 that examined investor behavior on the online investment platform eToro. On this social network, participants are able to view the stock market decisions of other investors and copy their actions one to one. It was interesting to see in the analysis what separated the successful from the less successful investors— namely, the versatility of their copied investment strategies. The more investors deviated from their own and others' views, the more money they earned. The best method was from those who integrated into their portfolio eight to ten widely differing market strategies from other investors. These portfolios were 30 percent more profitable than those of investors who predominantly copied market strategies from others who shared similar opinions to themselves.[19] It was possible to break up these echo chambers, however, by telling the less successful investors that diversity is what brings success. When some investors then changed their strategies accordingly, they saw their profits go up by 50 percent.

Never forget: although thought patterns and biases might seem very stable, immovable, and plausible—they aren't. A bias is not something that can be found in the environment; it is not a law of nature. It is rather a crutch that we use to make life easier for ourselves. This also implies that we don't always have to be right. Just as we stubbornly rely on our own stereotypes, the others around us have their own set of thought patterns as well. It can thus be very helpful for all of us if we often deliberately try other perspectives on for size. Peek into a Facebook group that has a different set of opinions than you do. Check out another newspaper or online news site that is written from a different political stance than your own. The more you expose yourself to others' points of view, the more well-founded your own views become.

Naturally, a discussion with others who have contrary opinions is not always fun, but it is useful. Several of my friends see the world in a very different way than I do. This often leads to heated debates, but I am always able to take a counterintuitive idea away from the discussion. I already know my own opinions; no one needs to explain those to me. This is also the reason why I am writing this book. I could just make it easy for myself and show you all the places where the brain is making its mistakes and how best to eradicate them. But that would only be half of the truth about the brain's weaknesses. Instead in each chapter, I also choose to emphasize the good things about the brain's fallacies. Even biases have their good sides—if you use them correctly.

Now you've learned it's no surprise:
We all see the world through biased eyes,
But the best way to learn if they're right or wrong
Is by switching them out and trying others on.

12

MOTIVATION

*How Our Inner Deadbeat Holds Us Back—
and How We Can Inspire Others*

W HEN I WENT up to receive my third grade report card
in 1993, I learned with startling clarity why motivation doesn't work. It was the first year in school in
which we were given grades, and a group of us boys were
eager to outdo one another (yes, yes, I'll admit it, we were little nerds back then). Who was going to make the cut? In the
end, we had all worked really hard. Each student was called
up to proudly receive their report card. So far, so good. Until
Daniela was called up and received a heap of effusive praise
from our teacher for having achieved the highest grades in
the whole class. Of course, she had most likely earned it, and

the teacher's praise was well intentioned—but, at the time, we didn't think it was all that great. And then to be beaten by a girl? Eeewww!

Don't worry. I'm not really suffering from some kind of elementary school trauma that I am trying to work out here and now. But something else became very clear: a teacher who elevates a student is eager to motivate that student. But in doing so, they create one victor and twenty-nine losers. Instead of motivating all of the students, they demotivate most of the class. It is not maximum performance that is being promoted, only competitive thinking, and as you will learn in this chapter, this is harmful especially to women's performance.

Motivational traps like this are everywhere: whoever performs best wins a prize; you can earn a bonus at the end of your project; let's reward ourselves with a piece of chocolate for mastering a difficult challenge. But, unfortunately, our brains aren't set up to be sustainably egged on by the chance of a reward. If high bonuses automatically led to the best performances, we would never have experienced the 2008 financial crisis.

At the same time, motivation seems to be a very important trait for us nowadays. There's always someone somewhere who needs to get motivated: students by their teachers, employees by their bosses, athletes by their trainers, ourselves by ourselves. It would be easy to think that without motivation we would be lackluster beings, sitting around apathetically with no drive whatsoever. And this is true. Because motivation is the mental impetus of our brain. It's our brain's kick in the pants that we call the ventral striatum.

Obviously, motivation is necessary, since it's not very easy to get ourselves moving. Three out of four people intend to make changes at the start of every year. The top resolutions are: live healthier, get more exercise, smoke less, spend more time with friends and family. The only thing missing from the list is the resolution to stick with one's goals. Because half of these people give up on their resolutions after only three months and there is once again a lot of space at the gym. We are always in a bit of a struggle with our inner deadbeat. We find it hard to stir ourselves to action and end up putting things off until the deadline is looming over our heads. How great it would be if we could rally ourselves right then and there and use tricks to fire up motivation in ourselves or others.

But why does this "deadbeat" have such a bad image? Obviously, we know that someone who lounges around, unable to push themselves to go to the gym or to finish their degree is never going to get ahead in life. But behind this behavior lies the aforementioned biological principle: we are simply not motivated by external incentives. We don't naturally run from one reward to the next and are for this reason not tamable as animals are. The fact that we are sometimes unmotivated is merely the price we pay for not being dependent on external rewards or praise. And it is only because of this that we are able to behave freely and independently. This is important because long-term motivation always comes from within.

This sounds all well and good—sometimes there are advantages to winning the fight against our inner deadbeat. But why is it so hard? What motivates us or others? What is, in fact, this motivation that we are all looking for, and where is it hiding out in our brain?

The problem of being Santa Claus

WHEN DO YOU feel particularly motivated? The prospect of a big reward isn't too bad. Such a reward can originate from vastly different quarters: externally (for example, if you are promised a higher bonus) or internally (if something is fun for you—for example, reading a book). Everyone is driven by something—and this works best when the reward is a little bit bigger than we expected. It isn't the reward itself that really motivates us—rather it is the hope of receiving just a little bit more. Or if we have once before been positively surprised, we can hope to be surprised once again. But you'll notice that this becomes more and more difficult to sustain, since it is quite a hard thing to orchestrate being surprised. It's like trying to tickle yourself. Fortunately, our brains can do it relatively well. The trick is this: keep your expectations low. That's the best road to happiness. Because if you don't expect anything, you can't be disappointed. Or to put it another way: someone who wants to do everything perfectly or always only wants the best will never be satisfied.

You have probably seen this in your own life. If you want to see a model for rewarding people poorly, you don't have to look much further than Christmastime traditions. Every year at a specific time on a specific day, gifts are distributed in a highly ceremonial ritual according to wish lists composed the month earlier by eager children. Expectations are high while the surprise factor is pretty low. Fortunately, this failure can be attributed to Santa Claus. After opening gifts one year, a four-year-old relative asked me: "I got this from Santa Claus. But what do you have for me?" If you leave the boring gifts to Santa Claus, you have the possibility of shining much more brightly in your gift choices later on. Santa Claus may have the most thankless job in the world. A lot of people think that he goes around with his sack of toys sparking worldwide joy and happiness in the eyes of children. But I am pretty certain that he has the hardest job of all, and it's not much fun. No one else has the potential of so easily failing to meet children's expectations. So, it's no wonder that he only ventures out to withstand this torture once a year.

Unfortunately, a lot of employee bonuses work in much the same way. First, you create an expectation that is difficult to fulfill. We even often forget about the amount of the bonus. The important thing is the element of surprise. For example, when your partner suddenly and unexpectedly brings home a bouquet of flowers, a box of chocolates, or some kind of elegant piece of jewelry out of the blue, this has a much greater effect than if they had waited for an anniversary or birthday. This is also a good strategy for saving money, since the surprise alone helps to strongly activate the reward centers in the brain.

The prediction error

MOTIVATION STARTS WITH a mistake in the brain. This sounds worse than it is, since I don't mean to say that we are doing something wrong whenever we feel a driving urge to do something. Instead, the neurons in our brain misjudge each other, and this is why there is any urge in the first place.

Whenever we are positively surprised, a moment of happiness flashes in our brain. Dopamine is released into our reward center, four times as much as usual. And this packs a punch, which is what we call the "kick." Rewards and dopamine go hand in hand. This is what we often hear, but this is only part of the picture. Because a shot of dopamine is not sufficient to help us feel rewarded and motivated. Our expectations also play a role.

Our brain has a region that is actually responsible for the reward sensation. It is the previously mentioned nucleus accumbens. This cluster of neurons in the limbic system (which is about as large as a dice) is the central station for a good mood—in fact it is a one-stop shop. If you read headlines stating, "Delicious food stimulates the same brain regions as good sex," or "A good book activates the brain in the same way as addictive drugs," this is simply because there is only one single region in the brain for this reward sensation. Good food doesn't necessarily have anything to do with a good lay (even though *accumbens* literally means "to lie with someone").

As I mentioned earlier, it is not enough for the nucleus accumbens to simply pump us full of dopamine in order to get our motivation up and running. Equally important is the moment of surprise, or expectation. This is controlled by our midbrain, or more precisely by the ventral tegmental area (VTA) of the midbrain. This region is located approximately where the spinal cord ends and the brain begins and contains the neurons that fire the dopamine to the brain's reward centers. In order to do this, the neural fibers stretch for several inches from the midbrain into the limbic system to the nucleus accumbens. The midbrain is thus constantly at work firing dopamine, depending on our expectations. If we expect a big reward, the amount of dopamine fired off is high. If our expectations are low, the amount fired off is also low. This dopamine-based activity is largely the hurdle over which a reward must jump. The higher the bar is set (the more dopamine activity is already underway), the harder it is for us to be surprised.

But sometimes it happens anyway, and this is usually because the midbrain made an earlier mistake by not releasing enough dopamine to correspond with the potential reward. Suddenly the dopamine release is much higher and this *difference*, this increase in dopamine, is what we experience as a reward. So basically, something that is an error in the midbrain is what makes us receptive to the feeling of reward. This is why this model is called "reward prediction error" by scientists.[1] Or, in other words, if our brains never made a mistake and were always able to correctly anticipate a reward, we would never be happy.

Built-in motivation

MOTIVATION IS NOTHING special to the brain. It's commonplace. Actions, movements, decisions—all of these are controlled by our motivation system. In principle, we are always motivated. We want to show what we can do, we want to be valued, we want to improve ourselves. No one really wants to lounge on the couch for all eternity. Everyone wants a goal they can be excited about. We enter the world already set to this basic mode—with the will to develop ourselves further. This is why it is so easy to inspire children and get them excited. When my neighbor's son was one year old, he one day used a seemingly unobserved moment to pull himself up on the side of the sandbox and let go of the rim, standing by himself on two legs. His face broke into a beaming smile, and he let off such a strong whoop that gravity quickly regained control over his diaper. Since that moment, he has never sat still again but always has to be standing up. This is motivation *in action*. Small children are permanently curious and want to conquer the world. They are overly excited when they are able to succeed in something. At least, I've never seen a three-year-old couch potato who argues, "Ah, just let me be, I don't feel like doing anything. And anyway, what do I get for standing up? Life is too stressful, it's all too much." For small children, it's almost never too much. They are always on the lookout for something new. And in this basic mode, kindergarteners are not much different than adults.

So how can you motivate people? You can't. It's impossible. You cannot motivate anyone, no matter how much you yell, "Just do it!" After all, you can't make someone hunger or thirst after something. Motivation is what happens when you anticipate affirmation for yourself and your performance. All you have to do is wait until the motivation comes on its own. Just as you get hungry if you have not eaten for a long time and suddenly have the urge to eat pizza.

If this is the case and motivation really is "built in," then why are we always complaining about being "unmotivated"? Why do we feel so sluggish about working, or going to the fitness studio, or studying our foreign language vocabulary flash cards? What happened to our inner drive in these cases, when it is supposedly so powerful?

The answer is that our motivational system has three weaknesses. Firstly, it wants to get as much for itself as possible. Secondly, it wants immediate gratification—it doesn't like waiting until later for its reward. Both of these often result in us putting things off and giving in to our inner deadbeat. And thirdly, the reward has to be something we can envision and something that has personal significance to us. This last reason is why we are surrounded by systems of demotivation that threaten to bury our inner drive.

When we don't care about ourselves

WE VALUE AN immediate reward much more highly than a future one. Our inner deadbeat uses this particular temporal uncertainty to squash our built-in motivation. In a sense, the deadbeat demotivates us indirectly by making sure that we value future rewards less than we value immediate ones. If, for example, we have decided to exercise, it whispers to us: "The couch sure is looking comfy right now." And even though we know that our deadlines are creeping closer and closer, he assures us with: "Tomorrow is another day." Of course, to say that this is "deadbeat-driven-behavior" might sound a bit harsh, so we call it procrastination, which means to put things off.

Procrastination is a prime example of the criteria that can undermine our motivational system. One weakness, in particular, makes us susceptible to putting things off, and this is that we feel relatively indifferent toward our future. This is not anything new. But an experiment conducted in 2008 was able to prove it. Test participants were told to taste a disgusting fluid "for the benefit of science" (the fluid was a mixture of ketchup and soy sauce). Afterwards, the participants were asked how much of the same mixture they would be willing to drink right away and then a few weeks in the future. And lo and behold: the further in the future their task was, the more the participants dared to commit to[2]—namely, a half cup. However, if they had to drink the unappetizing fluid again immediately, they were only willing to drink a maximum of two teaspoons.

Our future self does not have it easy. We don't spare any thought to loading up our future selves with tasks and burdensome obligations. Brain scans even show that when we imagine our future selves, we are not activating brain regions responsible for self-awareness but rather those regions related to thinking about other people.[3] No wonder we can't throw ourselves into a heavy round of CrossFit or we put off finishing our final thesis until the very last second. Our motivational system only pushes us to act when we ourselves are truly going to profit from it—and not some unknown stranger far off in our future. After all, we barely know our future selves.

What this implies, however, is that the more we know about our future selves, the more we allow ourselves to get revved up over long-term decisions. This might explain why some American retirement insurance companies offer a service in which they digitally alter a photograph of the potential client to show them what they will look like in thirty years. This is supposed to motivate people to purchase insurance. As long as it doesn't backfire. I can imagine, for example, that if this service is offered in Southern California, people might not run out to the nearest retirement insurance agency but rather to a plastic surgeon. "Look at this! I am going to look this terrible in thirty years! Help! I'll take the full Botox package, please." Oops, there went the retirement fund.

Where our inner deadbeat hangs out

THE SECOND WEAKNESS in our self-motivation system is that a reward in the future is not as tangible as a reward in the present. This temporal weakness of our reward system can easily lead to real consequences. Let's imagine that you were given the option to receive thirty dollars right now or fifty dollars in half a year. What would you decide? Most people would go for the smaller but immediately available cash. Which is ridiculous, since we would get more money if we would only wait for a little while. But men are particularly prone to this behavior, especially if they have just been looking at photos of attractive women. Interestingly, they are not very influenced by looking at pictures of sleek sports cars. Women, however, were more influenced by the photos of attractive sports cars than photos of attractive men (though this finding was in itself almost insignificant).[4] The good news for any guy who has invested in a Porsche 911 or BMW Z4-Cabrio: It works! Your strategy with the car and babes really works! Even the biggest slob can make himself seem interesting if he's sitting behind the right steering wheel.

Enough with the wisecracks—let me return to the topic. The reason for this time-based undervaluing of rewards has to do with how our motivational system is wired. Immediate rewards activate our nucleus accumbens more than rewards that are off in the future. In other words: better a bird in the hand than two in the bush. Better chill out a little now than to have six-pack abs sometime later on. This is not so much a lack of motivation as it is a stronger motivation for the immediate reward, even if this reward does not last in the long term. This I-want-it-now impulse is especially strong in our younger years. As we age, it grows weaker.[5]

Anyone who would like to investigate this effect for themselves need only put a piece of candy in front of a four-year-old and tell him that he may either choose to eat it now or wait for a while and then also get a second piece. This experiment is pretty mean to do on small children because even when they summon all of the self-control in their brains, very few of them are actually able to reign in their impulsive nucleus accumbens. This is not a new test. It was first conducted in the 1970s by Walter Mischel using marshmallows to tempt young children.[6]

It was interesting to follow what the various children struggled with years after the experiment. Incredibly, those who had been able to hold back during the experiment were more successful, had on average higher incomes, were better educated, and were less likely to be criminals.[7] Thus, anyone who is able to hold back now in order to receive a future reward has a psychological advantage.

But this is only half of the story. There is one important factor that is often left out when people refer to this marshmallow test: the children who participated in the experiment came from a privileged circle of society (they were largely the children of professors and scientists at Stanford University, one of the top institutions for intellectual elites), and they thus had good reason to believe in the promise of the second marshmallow offered by the researchers. When the experiment was repeated in 2016, results showed that the exact same neural mechanisms that helped privileged children to hold back and wait for a later marshmallow influenced children from poorer populations to immediately swallow the first marshmallows.[8] It was similar neural activity that resulted in different behavior. In the end, we must be able to afford the luxury of holding back. You don't go from being a dishwasher to a millionaire by optimistically (and possibly naively) holding back, but rather by utilizing every possibility.

In short, our motivational system aims to be rewarded as quickly as possible. For this reason, it is sometimes overly motivated and tempts us to procrastinate, since the present moment seems to offer more of a reward than a cumbersome task does. What helps is to think of the endgame. The study with the photographically aged participants shows that we are indeed able to abandon our impulsive short-term thinking if we can envision a future version of ourselves (even if this version is only a photo montage). The more clearly we are able to picture the long-term consequences of our actions, the more willing we are to face the unpleasant present actions. Be sure to actually carry them out! It is more important to do something than to do it perfectly. The brain only generates a reward impulse if there is something to reward. Therefore, during your arduous task it is best to concentrate concretely on the next step, even if it is very small. Working for a fifteen-minute stretch is better than doing nothing the whole time. And we should not constantly be picturing how much there is left to do. Instead, we should affirm to ourselves with each step how much we have already accomplished. This generates a short but valuable affirmation that propels us to take the next step. Our appetite comes with eating, and sustainable motivation comes only as we move along, step by step.

Our omnipresent demotivation systems—and how to fight them

I WOULD LIKE to be certain, at this point, that I am not making our inner deadbeat out to be stronger than it really is. Because, in truth, we are constantly looking for the source of our next reward and we are permanently motivated. And just as often as we tend to procrastinate, we also tend to pre-crastinate—that is, to complete a task immediately. We see this when people are given a choice to do something now or later. When participants are given the chance to pick up one of two weights and carry them to a point a few yards away, almost all of the participants will grab the closer weight even though they have to carry it a further distance than the more remote weight.[9] They often argue that they wanted to complete the task as quickly as possible—we apparently will choose some additional work if it seems to offer a feeling of being relieved.

Motivation is anchored into us and even if we put things off or laze around at the corner of the room, we are strictly speaking not unmotivated but rather fixated on the wrong reward (leisure). Instead, the problem is that we are surrounded by demotivation systems. This starts during our school years and continues through to our careers. At which point, everyone is trying out all kinds of new motivational tricks to give ourselves a new push. But eradicating demotivation is of much more importance than inspiring ourselves to new motivation. Thus, I would like to outline the "Top 3" motivation killers so that you are better equipped to recognize and get rid of them.

Demotivation system #1: Individual instead of group incentivizing

Just as I experienced in my school class that special praise for one student demotivated other students, many businesses practice a similar tactic with their employees. Instead of encouraging the entire team, an individual employee is singled out. This destroys the team. The greater the difference in rewards for individual team members, the worse the performance of the team as a whole.

This demotivational effect came to light in a study at the end of the 1990s in which researchers looked at the performance of athletic teams. They specifically studied baseball teams, which was helpful in this case because they offer a good mix of individual performance (for example, home runs) as well as team results (games won). The greater the income differences between the team players with the highest and lowest salaries, the fewer points they scored and the more games they lost.[10]

Individual motivation has the effect of demotivating the rest because if financial reward is limited, one man's gain is the others' loss.

In company call centers, rankings are frequently posted for the person who has called the most customers, sometimes with a bonus attached. Who has filled out the most customer contracts? Who has transported the most packages? At the end of the month, companies often post an "Employee of the Month" picture on the wall as evidence of how bad the others all were. But studies are clear that creating competition within a team, company, or group makes for a bad environment. Especially if the situation involves mental performance such as creativity. When researchers gave a group of people a concrete creative task (creating word puzzles in this particular study), individual performance was especially poor whenever only the best participant was rewarded for their work.[11] Because by doing so, one generates a sense of competition that compromises a portion of mental resources. Suddenly, it is no longer enough to simply do one's best, but instead you have to try to be better than everyone else. Women's performance level, in particular, is impaired by competitive thinking—their creative performances sank remarkably whenever they were faced with competitive scenarios within their team. Better performance is promoted by not offering any reward at all because then teammates are not comparing themselves with one another.

The remedy: Utilize the upside of group pressure

People desire social recognition. As much as we enjoy advancing as individuals, we are also social beings and are more motivated within a group than in isolation. Unfortunately, many reward systems continue to reward individual performance

independently from one's colleagues. You earn a bonus if you
have fulfilled a desired outcome regardless of whether or not
you cooperated with or exploited others. Many companies are
cautious about rewarding group dynamics since they are not
very clear and easy to control (and "soft" topics such as a company culture or atmosphere are not very compatible with the
measurability standards of many CEOs). It is much simpler to
merely measure and rate individual employees. However, team
performance is often much more significant than an individual's achievements. And besides, there are very few things that
are as stimulating as the behavior of our fellow human beings.
Peer pressure is an enormously powerful motivator. For example, if you have just finished watching a play and the rest of
the audience jumps from their seats and applauds loudly, it will
feel very hard for you to hold back and remain seated without
clapping. The same is true in the digital world. In the face of
the 2010 congressional election in the United States, a study
was conducted to try to spur more Facebook users to turn out
for the vote. Sixty-one million people on Facebook were sent
an appeal to vote. One group received only a message that it
was now time for citizens to vote again. The other group was
also shown photos from Facebook friends who had already cast
their ballots. It was no surprise, then, that people in the second
group were more frequently influenced to vote.[12] This was determined by having those surveyed click on an "I voted" button
once they had cast their votes and thereafter by determining the
turnout of the electoral districts under review. This only applied,
however, if the pictured friends in question were in fact "real"
friends, and not some vague Facebook acquaintances. Peer pressure is thus only effective if it takes place within a group that is
well known to us.

It is often not all that important that we are able to motivate ourselves but rather that we have an environment that is able to promote and constantly reinforce our motivation. For example, when we receive positive feedback from those people who are truly important to us (friends, family, well-known acquaintances, or colleagues), the feedback serves to activate us much more than if our boss pays out our long-awaited bonus at some point. This exceedingly more effective form of social motivation works much differently than one might expect. It isn't the individual reward that pushes a person but, instead, that one is able to give something to the group! In 2011, a campaign was started in Graubünden, Switzerland, to reduce household energy consumption. Instead of offering individual clients a simple bonus program in which they could collect premium points for each kilowatt hour they saved, organizers relied on the power of the community as a whole. They set up an online platform on which clients could share their energy consumption details. In addition (and this was very clever), they could invite their neighbors to join the energy network. This led to the townspeople forming small neighborhood groups that worked together to save energy. The more efficiently someone was able to cut down on energy consumption, the more bonus points were gained—for one's neighbors! Groups, and not individuals, were rewarded in this system. The result was that energy consumption went down by 17 percent, twice as much as in previous energy-saving campaigns.[13]

Why did some people choose not to take advantage of this? It wasn't possible in this system to be a freeloader and simply hope that the others around you would be energy conscious enough to earn you points. But this is where it's possible to really see group motivational power at work. Neighbors checked on one another (it was possible to see how much energy each neighbor was saving) but without an authoritative, hierarchical structure looming overhead. Instead, the dynamic became one of cooperation, in which neighbors helped and motivated one another. This idea of a motivational environment is diametrically opposed to the commonly practiced reward structure of our business culture. Instead of undermining the classic reward system of motivation for individuals, such a social system creates a self-perpetuating motivational group, strengthens social cohesion, and sidesteps mistrust and jealousy (the frequent side effects of classical bonus and reward systems in work environments).

This makes it possible to see how people really work: we do not flourish as trained machines but as individual members of a social group. There's a good reason that it is hard for us to be motivated by the usual motivational tricks and systems of reward—namely, that we like to solve problems in our own way, as well as to tackle them together in a group. Most bonus systems underestimate this principle value of human motivation and treat people as machines that can be lured by the prospect of reward into performing better and faster. Except that we are humans, and we don't run twice as fast even if we have twice as much fuel in our tanks. Especially when it comes to extraordinary mental undertakings, social cooperation, creative thinking, and problem solving, all of which cannot simply be motivated by offering some kind of incentive. However, we *can* remove factors that serve to demotivate and instead create a safe environment, a clever group dynamic (peer pressure), and take care to praise individuals (one-to-one, not in front of others). Because the ultimate reward is recognition and respect from one another. Neurologists don't have to take brain scans to be able to observe this dynamic, though they still enjoy doing so.

Demotivation system #2: Attaching reward to performance

There are many who believe that rewarding someone for a great performance is a good motivational tool to help them repeat their behavior. But this is only true at first glance. Because people actually don't want their performance to be rewarded. They, instead, want their individual selves to be rewarded *for* their performance. Because if it is only about rewarding a particular result, the person doing the result is replaceable. It isn't the person themself who is important but the performance or result.

A study from 2010 was even able to measure this outcome. Study participants were asked to solve a simple cognitive exercise in the lab: clicking on a stopwatch exactly every five seconds. At the end of the exercise, they were rewarded for their precision. If they clicked within fifty milliseconds of five seconds, they received the equivalent of about $2.20. Those participants who were promised a reward were naturally energized and made more of an effort than those who were not rewarded since financial incentive can certainly influence one's drive. But as soon as the financial reward was suspended, the participants who had been fixated on winning money were suddenly no lon ger interested in continuing the task. Brain scanners were able to show that the reward centers of the brain that generate our inner drive were less active in the rewarded participants once the financial reward was removed than they were in those who had never received any reward in the first place.[14]

In other words: by giving people a financial incentive, they were fixated on the concrete reward and not on the task. As soon as the reward fell away, the basic inner drive, our "built-in motivation," was no longer there but had been buried by the lure of money (this is why neuroscientists call this the "undermining effect"). "So what?" you might say. "Does it really matter whether someone does something for money, as long as they perform well?" Well, maybe not, as long as the task is a simple one such as clicking on a stopwatch. But over the long term, people become dependent on this manner of financial reward. The result is that an employer has to pay financial bonuses so that their employees feel valued, the way that a valued employee without a bonus already feels. Why not just save yourself the money?

The remedy: Spark people's ambition

People want to improve and show what they can do. In spite of the inner deadbeat, people still go to the fitness studio (they even pay for it), play music in their free time, and look for the newest recipes in order to improve their skills. In the U.S., more than 70 percent of all firefighters are volunteers. And all of this without any financial compensation. Why in the world do people choose to do this? Because affirmation of one's service is more valuable than money. Anyone who is finally able to play Beethoven's Piano Sonata no. 9 after long years of practice or saves a child from a burning house or is able to bake a meringue dessert without it collapsing does not need any extra reward for their accomplishment. The affirmation of one's own ability is completely sufficient. This is even true in professional business because the greatest successes cannot be bought: finishing a marathon at the Olympic Games, playing a concert in front of a sold-out concert hall, writing a bestseller. You can pour as much money as you want into the effort, but if the motivation does not come from inside yourself, you won't get very far. What really motivates people (long-term, that is) is the search for personal advancement and for affirmation. No one is going to remember the masterpiece stroke of business by an investment banker in fifty years, even if the bonus was really high. But they will remember the invention of the iPhone, the first Mercedes, or the invention of the printing press. Of course, most inventors also naturally thought about financial gain. But you can only achieve this if you are pursuing a more important goal: proving it to yourself.

Demotivation system #3: Rewards for mental activities

vnvnvnvnvnvnvnvnvnvnvnvnvnvnvnvnvnvnvnvn

Oh, excuse me. I was simply performing a type of exercise that can easily be driven by reward: monotonous motor activities. Simple mental actions that even a machine could complete. And this is precisely the point. It is when we are required to work automatically that rewarding people really starts to pay off. Because who really wants to type out two letters over and over again? It would drive anyone crazy, and the only people who could stand it are those who are rewarded financially. Participants in a study were told they would receive $15 for typing the letters vn six hundred times every four minutes. Those who were really lucky were in the second group that was offered $150 for the same amount of typing. It's no wonder, then, that the potential rewards drove participants to type as fast as they could. Their brains switched off, they simply hammered away. Simple tactic, enormous results. But this doesn't work very well if we have to think actively. When participants were told to complete simple arithmetic problems for a chance to earn a high reward of $150 per every ten completed problems in four minutes, their performance sank (compared to the low-reward control group).[15]

To put it bluntly, when it comes to cognitive tasks (math, language, organization, creativity), rewards destroy our performance ability. The higher the reward, the worse the result. For example, school lessons. It would seem that learning in a classroom is best when students are required to complete their lessons, and when the best students are rewarded. A lot of parents do this too, by giving their children a financial bonus for their grades: ten dollars for every B, twenty dollars for every A. (Don't do it! It is harmful!) When study participants were asked to memorize vocabulary words for a test, their performance depended on whether or not they were rewarded. After a week, those who were rewarded with one dollar per correct word in the previously administered test were able to recall fewer words than those who were not rewarded.[16] As soon as a reward enters the picture, it is no longer about doing one's best but rather about getting the reward. If you are routinely rewarded with fifty dollars for your report card, going to school becomes a hard way of earning the fifty bucks. It shouldn't come as a surprise when the student completely loses their inner drive later on.

The remedy: Unleash mental capacities

People want to act independently. It isn't enough only to have the prospect of personal advancement, of development and improvement, or the feeling of being recognized for one's performance. No one merely desires to do somebody's groundwork without having the chance to show one's own individual skills. Only those who feel that they are self-determined in their actions will be able to reap a sustainable feeling of happiness in their accomplishments. You see, our brain is very specific at making the distinction between something we have accomplished by our own efforts and what simply came to us. It is only when we feel that we have really "earned" our accomplishment that our reward center continues firing.[17] If, on the other hand, we are successful by chance without having had to do anything, the reward is less fulfilling. In other words, winning a million bucks in the lottery does not feel as amazing as earning a million bucks. Instead of trying to encourage employees to reach a particular outcome by offering financial incentives, companies should try every now and then to have faith in their employees' own motivation by, for example, telling them, "You have two free days to do whatever you want for the company. Just report back later on what you were able to come up with."

The reward paradox

SO, WE KNOW that our motivational system has a weakness for quick and immediate praise and is easily undermined by misplaced rewards. This might sound like a disadvantage at first glance, but what it really shows is that we don't function like robots, and that we are instead flexible and unique in our manner of thinking.

Each of the aforementioned demotivational traps makes the same mistake: they treat our brain as if it was a machine. As if they were saying to us, "If you deliver this performance, you will be rewarded." But this only works if it is a simple, automated type of task that does not require much thought. We are rewarded for the results of our labor but not for the reason that we undertook it. The underlying motivation is completely irrelevant as long as the desired results have been reached. If you offer a reward for something, you are encouraging the motivational system to find the most efficient way of earning the reward. This can quickly backfire. When the first railroad tracks were laid in the U.S. in the mid-nineteenth century, those responsible tried to keep costs down by linking them to the built track: $50,000 was allocated per track mile (an immense amount of money in those days). Then Thomas Durant, the project manager for Union Pacific Railroad, had the idea: "Why not simply draw a few extra yards into the tracks?" And thus the planners drew a few extra unnecessary curves into the plains of Nebraska. IBM made the same mistake when they promised to pay their programmers for every line of code. The result was that the programmers simply wrote in a couple of extra lines of superfluous code and earned themselves a bonus. The same thing happens in avoiding penalties. When, due to the terrible smog, Mexico City enacted new rules stating that drivers could only use their cars on certain days in the city, it turned out to be a blessing for the auto industry. Because anyone who could afford to do so simply bought an extra car so they could drive around twice as often.

It is thus good to note that rewards drive us—to earn the reward. But they don't necessarily activate our inner motivational system. If the reward is taken away, our inner drive is extinguished. As bad as this sounds, it is actually a good thing because it shows that we are more than circus ponies. We don't allow ourselves to be backed into a static if-A-then-B mode of thinking. We are not so easily motivated by a bonus because we are individuals and therefore desire to be personally valued. The greatest achievements in history have never come about because of a promised financial bonus. Of course, it's nice to be paid, but in most cases, this is not the goal. It is much more important for people to be valued as self-determined individuals for their service. This might sound cliché but it's true.

13

CREATIVITY

Why We Can't Be Innovative at the Push of a Button—Yet We Always Have New Thoughts

L ET US TURN now to the supreme discipline of the human brain. Making a few decisions, learning about something, or juggling a couple of numbers is all well and good. But nothing can compare to the high art of human creativity. This distinctive form of creativity is something that only we can boast—and that includes you. And that is why I am going to give you some free space for your creative power. Pen ready, your task is to illustrate in a creative way the word "enjoy." Because performance is measured as the rate of work per time unit, you only have thirty seconds to complete this task. Here goes:

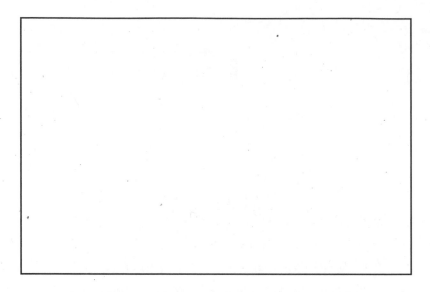

"Wait a second," you might be saying. "Not so fast. How am I supposed to find a pen so quickly? I am here to read this book. And anyway, how am I supposed to be creative at the push of a button? I need more time!" This brings me to my key point: we are unable to use our creative powers very well when we are under pressure. In fact, the more stress we are experiencing, the more constrained we feel, which affects our actions and behavior.

You probably already know this from constructed creativity tasks as well as from your real life. For example, if you have been studying for an important final exam. You were up all night for weeks trying to imprint the last vital details of the test material between your ears. Then the day of the test finally arrives, it's all or nothing and, in a few minutes, your fate will be decided. You are tense to the tips of your fingers and toes, and the pressure is enormous. If you are feeling good, you will be able to regurgitate the learned materials on cue. But when you are faced with an open question, where the examiner says, "Now use your creativity! Here are three colored pens, please create something new with them!" you suddenly run into a problem. Because it is very hard for us to cope with such ambiguous tasks under pressure.

We face these kinds of situations all the time, in which we are required to do something in a very short amount of time: "Please develop a crackerjack opening slide for the company presentation!" Or: "Honey, it's my birthday tomorrow, and I hope you are going to surprise me with something!" When we feel the pressure rising, we become particularly uncreative and tend to revert to our usual modes of behavior. Our method of choice is: don't think, just do.

At the same time, new ideas are what fuel our economy. And they are also, by the way, the only resource we have that won't one day run dry. Oil, coal, uranium, sand, water, even sunlight and time—everything else is limited. But ideas are not. They are thus the perfect raw material to use lavishly. This forces the question: If there are so many ideas to be had, why do we handle them so sparingly? Why don't we get the good ones in the exact moment when we could most use them? For example, when we feel under pressure? How often have you kicked yourself after a conversation for which you only thought of the perfect response two hours later? Why do we always find ourselves saying, "Of course! That's what I should have said!" instead of having the perfect punch line in the perfect moment?

Although they are infinite, new ideas are precious commodities. This is a paradox, and it comes down to the truth that original concepts are not manufactured on a factory line. We can neither plan, regulate, nor produce them. We can't even measure them. Or, how might you quantify a good idea? When do you consider a company especially innovative? When it has two thousand patents, perhaps? Or once a single idea has turned an entire market on its head?

It's no surprise that the whole world is always on the lookout for the next big thing—something radical, "disruptive," or "market-shattering," as they say in Silicon Valley. And yet, the creative thinking required for such ideas is hardly tamable. Our brain does not work as efficiently as we wish it would to produce ideas.

Even though it seems that our creative thinking leaves us in the lurch right when we need it most, this mental behavior is, in fact, an important process in our brain. The fact that we are unable to control our creativity is the price that we pay for being creative in the first place. Our idea processes may not be controllable or easily channeled, but this is the reason why they are still possible at all.

Measuring ideas

THE JOB OF a creativity researcher is not particularly easy. First, it is a very tricky task trying to come up with creative ideas in the lab. After all, you can hardly tell participants: "Now, be creative for a few minutes so I can measure what is happening in your brain!" And if that wasn't enough, creative ideas are very diverse and hard to translate into numbers. What makes a thought original or unique—is it when it is very "different"—or especially useful?

In order to study the lateral-thinking potential of our brain, researchers use tests that highlight the individual facets of our creative thinking. The process is not unlike an intelligence test that probes our mental abilities across various disciplines. While things like logic, memory, and language and math proficiency are important in an IQ test, other skills are also at play in these lateral-thinking tests.

For example, the skill of divergent thinking. The brain's ability to abandon well-trodden paths of thought or to jump out of mental boxes is unquestionably a very important characteristic of a creative brain. We have already seen (chapter 11—Biases) how our brain loves this kind of fixed thinking and routine. But these things won't help on a Torrance Test, which measures how willing we are to abandon our mental comfort zones in order to come up with new possible applications for everyday objects.

A Torrance Test could be something such as: coming up with as many possible uses for a toothbrush in three minutes! Alternatively, you could also substitute any other regular object to be repurposed, such as a piece of paper, a pillow, a plastic bag. The important thing is that the object must have a regular function from which the test taker will need to detach themselves during the test. At the end of the test, the creative ideas generated by the participant must be evaluated. Suggesting, for example, that a toothbrush can be used to clean a car's tire rims is not particularly resourceful (and earns therefore only one point). But if someone uses a toothbrush as a paintbrush, going even a step further and painting a naturalistic picture of an apple tree in the neighbor's garden, they earn three points. After three minutes, average creative participants earn around fifteen points. Theoretically, this could be achieved by either coming up with fifteen different possible applications for a toothbrush or five very divergent uses—complete with explanations of how each new function would be put to use. In each case, it is important to note that there is no right solution. There are only solutions that are more or less good. Which category the ideas fall into is judged subjectively at the end of the test. Because creativity is

never right or wrong, rather it is constantly breaking with hab-
its of thought. And this makes creativity the natural enemy of
efficiency.

Intelligence vs. creativity

RESEARCHERS HAVE BEEN administering the Torrance Test since
1966, constantly adjusting the test and environmental conditions so that the average score remains the same. The test is
thus standardized in the same way that IQ tests are readjusted
every few years in order to account for mental progress. The IQ
test is structured so that an average delineation point is 100.
Regardless of how many tasks are solved in the test, half of the
test takers score above 100 and the other half below.

Interestingly, it turns out that over time, we must correctly
answer more and more questions on an IQ test in order to reach
100. In other words: our mental intelligence increases by three
IQ points every ten years. If the test were not readjusted accordingly, a participant from 1960 with an IQ of 100 would score
only an 85 today—even if they solved the exact same number of
problems. This phenomenon is called the Flynn effect.

This problem does not apply to the Torrance Test, however. This particular test has had to be readjusted five times
since 1966—but instead of becoming more difficult, the test
has been made simpler over time because the creative abilities
of participants has decreased.[1] This implies that while we are
becoming more intelligent, we are, at the same time, becoming
less and less creative. This effect was especially significant in
children from kindergarten to third grade, which may indicate

that adults perform worse than children. And if you aren't the sharpest crayon in the box to begin with, there's not a lot of wiggle room left on the lower end of the spectrum.

Could it be that we are paying for our increased intelligence by becoming less and less creative? The more IQ tasks we are able to solve, the more our ideas seem to conform to norms. This is not at all surprising when you realize that intelligence means the quickest way to reach a previously known solution—and not the development of a completely new solution. Whereas in an IQ test you have to find correct answers as fast as possible, no IQ test calls for interesting open-ended questions. That's why an exercise such as "Try to create a new task to evaluate the concentration capacity of your neighbor" will never be part of an IQ test. Intelligence pairs up beautifully with our productivity- and efficiency-focused economy. Intelligent people solve problems quickly, in a straightforward manner and free of errors. Those who are intelligent are considered productive and mentally superior. This is good but, unfortunately, is also replaceable. Because if problems can be solved efficiently, they are too easy for us. The solution is already present in the intelligence test, and we just have to find it quickly. Computers will eventually be able to completely assume these responsibilities for us. It is futile for us to try to reach this level of intelligence against algorithms in IQ tests. Because sooner or later, we are going to lose.

Creativity, on the other hand, means thinking alternatively. And as much as everyone talks about wanting to be creative, innovative, and disruptive these days, the unbridled nature of creative thinking does not fit harmoniously into a business culture made up of metrics and spreadsheets that are meant to be controlled and optimized. Creativity does not mean doing everything correctly. Creative ideas are always somewhat "false," erroneous, and uncommon because every new idea represents a break with a generally accepted way of thinking.

Therefore, before you want to become more creative, keep in mind that those who think outside the box can rub people the wrong way; they question established work patterns and don't merely accept them. They are not rule followers; they are rule breakers. They don't tend to get along with authority, are constantly trying out new methods, and are thereby destroying a lot of things along the way. They ask tough, uncomfortable questions, often offend others, and don't stick to fixed working hours. In brief: they are annoying. We don't tend to find this behavior to be a good thing, but it is essential. Because if you only ever follow accepted patterns, you will only generate predictable and boring ideas instead of new, provocative ones. This is why it is important to have the freedom to take detours and work inefficiently at times in order to arrive at a different location. On the other hand, only thinking out of the box and never in a straight line will make it hard to reach a goal. Ideas must also be applicable and useful at the end of the day. Our brain is thus tasked with holding and solving the necessary tension between thinking in ways that are off the wall and zany, as well as ways that are efficient and productive.

Thinking twice lasts longer

WHERE DO GOOD ideas come from in the brain? Anyone who is hoping to locate a "creativity region" in our brain (similar to our "language area" or "vision center") is in for disappointment.

Because creativity occurs through the fluctuation between focus and deviation. In general, our brain is constantly generating entire heaps of possible ideas, which it then has to filter out. You have already learned about both of the networks responsible for this task: the default mode network (from chapter 6 on daydreaming) and the control and decision-making network (from chapter 9). Good ideas thus come into being when the brain combines its ability to fall idle with its decision-making strengths.

If you stick someone who has just taken a Torrance Test into a brain scanner, you will see just how well these networks can collaborate.[2] If test participants are asked to think of alternative applications for bricks or newspapers, their default mode network suddenly sparks up. Let me give you a simple recap of chapter 6 so you don't have to flip all the way back: our default mode network is located from the middle all the way to the back portion of our brain and includes the regions that are responsible for mind wandering and daydreaming. This is very practical, because if you want to solve a problem with creative and divergent thinking (that is, thinking in all different directions), the ability to go back and forth and all over the place is very helpful.

Most of what we think is just junk. Fortunately, we don't even realize most of it because our brains sort out the really useless stuff. To do so, our default mode network coordinates with the control region of our frontal lobe that you learned is a part of our decision-making network. The insular cortex and cerebral cortex, especially, take on the task of sifting through possible ideas. While the default mode network works to free up space from all kinds of obligations, the control network filters the bad ideas out, and it does so relatively quickly. Brain scanners show that the connection of both networks is so strong after mere seconds that only the most valuable ideas are left. The decisive factor is this interplay between the two networks. For example, if test participants are required to think of only commonplace uses for everyday objects like a brick, the connection between these two networks is weaker.[3] The opposite is also true. The better these regions can cooperate and are linked through their neural connections, the more original are the ideas generated. For very creative people (who are able to think up unique uses for a toothbrush or a brick) taking the Torrance Test, studies find that their network connections are particularly robust.[4]

In conclusion, it is not the number of ideas that counts, but rather that a balance is established between the generation of new ideas and the ability to filter them out. In addition, tests have also shown that we are able to generate a higher number of ideas when we are more relaxed and even when we sometimes are not even concentrating on our task. Ideas sometimes seem to come out of nowhere and to strike us spontaneously. It is possible to study these kinds of "aha" moments in the laboratory. To do so, a different kind of test has to be employed: the "remote associates test" (RAT). Rather than measuring our divergent thinking, this test looks at the opposite, our convergent thinking, or how well we are able to single out the best ideas.

The art of not thinking

COMPLETE THE FOLLOWING task: Which added word meaningfully transforms the following three words into new words?

cream / skate / water

Okay, that was easy: ice. But what about this one?

hand / dish / opera

This is noticeably more difficult. It can take people up to half a minute to think of the solution. Take a bit of time to think about it before reading on. You've earned it.

Wait, don't just keep reading! Seriously, relax and think it over.

If you truly did take a moment to think about it, what was it that you actually did? Did you keep your eyes focused on the three words and try to combine them with other words? Or did you maybe let your gaze wander away and stare out into nothing? Interestingly, most people prefer the latter behavior and researchers can, in fact, measure that in looking around inconsequentially, we actually see nothing! Even though our eyes may be open, the visual centers of the brain are less active than when we are actively looking at something.[5] As if we were a bit blinded to the things around us. But where else should we look if not at our surroundings?

The explanation is that we are activating our default mode network and turning our gaze inward toward our wandering thoughts. We principally have two options for solving the above puzzle. We could proceed analytically—this is the more time-consuming, though easier, way. Analytical thinking is always available at the push of a button and, in this way, we could go through a whole list of words one by one to solve the puzzle. "Hand" might be completed with "wave" to be "hand-wave." But "wave-opera?" What is that? A new musical set in the

Pacific? That doesn't really make sense—so let's try something else. "Handmade" sounds right, but then "dishmade" does not. It's back to the drawing board...

This method is fine as long as you are able to think quickly and have a large vocabulary at your disposal. Analytical thinking is something for efficient mathematical geniuses. But who of us is a genius? If one is an algorithm, it's easy to solve a problem like this. But for normal human beings, it's a bit harder. And yet we are constantly trying to take the analytical approach because it is always available. Even the most uncreative person can pull themselves together and try combining a few words until something clicks. All you have to do is focus and don't allow yourself to be distracted and at some point, you'll find the solution. We are taught this in school and throughout our whole lives: whoever is focusing on their task is considered productive. Everyone else is a dreamer.

But is that true? There is, in fact, another way to solve the problem. Through insight (these kinds of word puzzles are called "insight experiments" in science). First, you study the words intensively, and then you allow your thoughts to wander. It sometimes happens that the answer "appears" to test takers all of a sudden. It might sound strange but it's true.

Although the purpose of this task differs from the Torrance Test, the basic underlying principle of thought is the same: first look for several different possible solutions and then select the best one. And both networks are equally necessary in this case too—the basic default mode to generate ideas and the control network to filter them out. The most important control post for new ideas is located in our frontal lobe and wraps itself around our limbic system like a belt (though it remains separate). It is here that it is decided whether an idea will make it from deep down in our default mode network up to the level of consciousness.[6]

This brain region is truly active in promoting alternative ideas and has a remarkable penchant for absurd thought. While our basic default network is going through the possibilities for the above word puzzle—why not try "-wave" or "-washer" or "-glove"—our frontal lobe is paying attention to the less prominent network activations—for example, the solution "soap." All of this takes place subconsciously. It is only when our selection centers have located the right answer that our control network is strongly activated, enabling our thoughts to rise up into our conscious thought.

Symbiosis of thought

IN ORDER TO have a creative thought, we must have both elements: the unconventional and the wacky, which will be further filtered and substantiated. The brain carries out both veins of thought in a dualistic interchange, a task for which it is surprisingly well situated. Because only this interchange system is able to meet both requirements for generating a creative idea. It's not enough to simply churn out a bunch of original thoughts (which the default mode network would be able to do on its own). New ideas must also be useful and applicable to solving a problem (which is managed by the control network).

Creativity is thus more than pure fantasy or an inventive effort. I could, of course, take a couple of different colors of paint, throw them up on the wall and call it art—but for me, this is not so much creativity as rubbish (unless someone is willing to pay a high price for my "painting"). Creativity is a method for solving problems that developed in order to break free of thought patterns when thinking fails to get us closer to a solution.

But when is an idea creative? There is only one criterion for this: when someone else says that it's creative. That's it. There is no measurable indicator by which to rate a new idea. There is only personal recognition from someone else. The application of the idea might, of course, prove to be useful (which can be measured), but the initial assessment of an idea always comes about through subjective feedback. This is why it is so difficult to build creative computers. Computers might be able to spit out a bunch of new "ideas," but the selection, the valuation, and the application of the ideas cannot be digitalized. At least, not yet.

In this respect, every single innovative idea is related to the arts. There is no such thing as an absolutely correct innovation; there is only always a subjective one. Someone has to say that an idea is good, otherwise it is destined for the landfill. Is, for example, the Internet a useful technical innovation? Absolutely, you might say. But if you are living far away from civilization, in the Australian outback, the Brazilian rainforest, or the Canadian Arctic, without grid or power connection, the Internet is a pretty useless concept. In this way, good ideas are like puzzle pieces: they always require a matching counterpart.

Good ideas are something else as well: rule breakers. Because creative ideas rarely adhere to norms and regulations—they change them. This is precisely the nature of divergent thinking, to find *alternative* applications. But in order to do this, the original rules of application, the mental boxes for how we are supposed to use a toothbrush or brick, have to be obliterated. The brain is only able to do this when it is not focused on the task but is rather allowed to wander around mentally (in its default mode network).

If we believe creative thinking is such a great thing, why does it often seem so hard for us to come up with new ideas?

The answer is that our dualistic system of thinking has two main weaknesses. The first is that our default mode network is prone to excessive concentration (for example, through stress or mindfulness training) and loses its ability to wander mentally. And secondly, the control network is heavily influenced by fear or threatening situations, which leads to it filtering out more information than it should.

Stress, a contrast regulator in the brain

STRESS IS A creativity killer par excellence. If you want to make certain that you won't be able to think creatively, put yourself into a stressful situation. The deadline will live up to its name and kill any new idea you might have.

In moments of high stress, our brain switches into a kind of crisis mode. After all, when the going gets tough, that is not the time to play around with your thoughts but rather the time to focus concretely on what lies before you. In order for this to work, the brain releases certain neurotransmitters that help us to be more alert and focused and, one could also say, narrow minded. Under normal conditions, these neurotransmitters act as direct messengers: one cell sends a transmitter straight to the next one, and then this one becomes activated. But this changes under stressful conditions. In this case, neurotransmitters (such as norepinephrine) are widely distributed as though by a watering can throughout the cerebrum, which has an effect on the default mode network. These neurotransmitters no longer have the task of transporting information from one cell to the next but are now tasked with indirectly controlling how other cells are stimulated. This is why these are referred to as "neuromodulators."

One way to imagine this effect is to think of a photo filter for Facebook or Instagram. For example, if I want to share a really terrific picture on the Internet, I need to have a cool motif. But that isn't enough. Because I can create the most intriguing images by manipulating a few special effects, such as more contrast, saturation, or a vintage look for the particular atmosphere. The informational contents of the photo don't change (the motif stays the same), but the effect is altered. The same photograph can be changed to look antique, funky, or psychedelic. This is the same thing that neuromodulators do in our brain. They don't change what we perceive; they change how we perceive it. In the case of norepinephrine, this perception becomes more focused and fragmented. When we are under stress, our neuromodulators function as a contrast regulator, suppressing unimportant background noise and emphasizing what is most vital.

Biochemical blinders

IN CASES OF acute stress, norepinephrine acts like a set of biochemical blinders. This has the obvious benefit of allowing our brain to focus quickly on details and to suppress any mental drifting.[7] Within this tunnel view, we are able to reach the highest level of concentration but at the cost of diminished creativity.

This can be fatal, as illustrated by a well-known anecdote from the history of firefighting in the U.S.[8] When a fire broke out on August 5, 1949, in the Helena National Forest in Montana, sixteen so-called "smoke jumpers" were sent into the forest to fight the fire. This special unit is dropped by parachute into a burning area and has received extra training to be able to battle fires in rough terrain on their own. In 1949, the problem arose with a dangerous combination of unfavorable weather and geography. The men had been dropped into a steep canyon with a 70 percent grade on both sides. With a temperature that day of 97 degrees, a stormy wind blew the flames toward the group, and suddenly they found themselves chased by the fire. They realized the danger too late and became gripped with panic. In their fear, the men began to run up the steep slopes, tossing off their heavy equipment and fleeing for their lives. But the fire was too fast. While the others scrambled away from the onrushing fire, the foreman, Wagner Dodge, realized that running was hopeless. He suddenly had an idea and lit a fire in the dried grass before him and began to run after it as it blew through the canyon. He lay down on the scorched ground and was protected from the oncoming blaze. Thirteen men perished in the fire. Two men were able to climb onto a rocky knoll and save themselves. Wagner Dodge survived because he was able to set aside his mental blinders for a moment. He later recounted that he had never before heard about the possibility of lighting a "backfire." This is difficult to verify retrospectively but, in any case, this story highlights how, under stress conditions, we tend to stick to our standard routines. And even when they don't work, we continue to stick to them nonetheless. This is why only one out of sixteen men was able to come up with a good idea under acute stress.

Most of us, however, are not threatened by surging fires in our everyday lives. But the basic principle remains the same: stress makes us narrow minded in a critical moment, and this concentration heightens our focus at the cost of our creativity. A similar effect occurs, by the way, through mindfulness training, which is very much in vogue at the moment, and which has as its objective the reduction of stress through practiced concentration. The paradox is, although following these methods does indeed reduce stress, mindfulness methods simultaneously work to undermine our default mode network. In insight experiments, for example, participants who were asked to draw connections between words were able to progress analytically following mindfulness or concentration exercises, but they never experienced an "aha" moment.[9] From this, we can conclude that concentration is a good thing if we need to select and refine one idea from a whole variety of ideas. But if we need to first gather several ideas, it's easy for us to fall into a concentration trap.

A good mood makes for good ideas

EVEN THE CONTROL network has its weaknesses. Sometimes it overshoots the target, especially if we are anxious or in a bad mood. In such cases, we lose our overview and instead try to avoid as many mistakes as we can by throwing ourselves into the details. It is similar to how we react under stress, but now there is another mechanism at work. When we are anxious or afraid, it is our control network and not our default mode network that becomes overactive.

What this means can be illustrated by the following experiment. Here you can see two different patterns:

```
o     o                    #
o     o          #              #
```

Which of the above patterns does the following resemble more?

```
          #     #
          #     #
```

Take a close look. There is no right or wrong answer to this problem, only a narrow definition (you are paying more attention to the character) and a broad definition (you are paying more attention to the pattern). Interestingly, it seems that sad people are more prone to pay attention to the characters. Happy people lend more significance to the overall pattern.[10] There's nothing wrong with this, but what we see is that creative ideas are most often generated when we are able to take a step back and look at the greater picture. And this works especially well if we are in a good mood.

As we learned in the chapter on decision-making, fear and avoidant behavior activate our insular cortex. This insular cortex is fittingly located at the exact intersection of the default mode network and the control network. In this way, our out-of-the-box input is sorted and filtered before it is released into our control network. We should therefore not be surprised when participants in creativity tests think more analytically and are less prone to free association the more negative their moods are. Participants who are encouraged to recall particularly depressing moments in their lives, instead of spending their time considering more positive hypothetical solutions, experience fewer "aha" moments when they are later given word-linking tests.[11] Hearing a joke, on the other hand, makes one more creative.[12]

Analytical and detail-focused thinking certainly has advantages if we are trying to avoid making mistakes in a critical situation. It makes sense that our control and decision-making networks restrict our filter mechanisms to keep us from getting distracted by farfetched ideas during a threatening situation. Creativity is a privilege or luxury that we must be able to afford. And this is particularly difficult if we are spending all of our resources on avoiding mistakes. A much better breeding ground for creativity can be had under safe conditions.

The price we must pay to be original

IN PRINCIPLE, THERE is only one single rule for creative thought: a creative thought must appeal to others—it has to be contagious. Nothing more, nothing less. But this is so terribly imprecise that our brain has developed more networks in order to cover the spectrum of creativity: some that are lavish on the one hand and others that serve to sort on the other. Only by reaching a balance are ideas able to arise.

The fact that we seem to lose our creative powers every now and then (when we are under pressure or feel afraid) is the price we have to pay for having this balanced system. Because we need both our filter and our daydream networks. If the first one stops working, this can lead to schizophrenia and hallucinations (which is in a sense an overriding and uncontrollable form of creativity). And without our daydream network, we would land in the accounting department. This is a joke, of course, since people can be creative in any department. Anyone who owns a business and has to fill out a tax return knows what I'm talking about.

There is thus no single rule for creative thinking. This is why computers often run up against a wall whenever they are faced with a problem requiring a creative solution. Intelligence? No problem! All you need is to be able to compute quickly. But whether or not an idea is innovative or total garbage is much harder to judge in advance.

For example, are you familiar with the smartphone app called Yo? Never heard of it? Then you're missing out. Because this app made it possible to send a "Yo" to all your friends and acquaintances.[13] That's it. A really short message consisting of a single word. This might sound totally absurd. So absurd, in fact, that the two Israeli developers anonymously put the app onto the market in 2014. They didn't want to look like creative losers. But three months later, investors were putting millions into the app's development, and they were cooperating with a Red Alert app that allowed people in Israel to warn one another during rocket attacks in the Gaza conflict. Currently, companies are experimenting with Yo to be able to communicate with their customers directly and without any chitchat whenever they have some exciting news. No one knows if anything will come of it, but this is precisely the point. No reasonable, thinking person, and especially no computer algorithm, would have spared Yo a second thought. But good ideas are often unreasonable, and no one knows what, if anything the Yo app will be able to accomplish in the future.

Just imagine that you had come up with a new business model for hotels and overnight stays in the mid-2000s. You sit down and analyze the field. Remember that these years are marked with uncertainty: terror attacks in Madrid and London, as well as in tourist locations across Asia. The U.S. extends its Patriot Act to intensify its fight against terrorism. Security is priority number one. But in 2007, three guys from San Francisco had the idea that people might want to privately rent their living rooms out to total strangers. In an era in which the U.S. had clamped down on its borders more than any other time in history, people would be expected to allow individuals they had never before seen into their house. Neither the hosts nor the guests could have any idea of what it would be like. So much for security. What a totally stupid idea! So stupid, in fact, that the company is worth over thirty billion dollars today, and its worth continues to skyrocket. Airbnb was founded in 2008 and works so swimmingly well that some countries and cities have had to rein it in with special laws. In retrospect, it all seems logical in light of the rapid evolution and success of social networks. But none of this was so obvious ten years ago. Today, algorithms work to optimize the Airbnb online platform, making its search and pay functions quick and efficient. But algorithms could never have predicted success from the counterintuitive start of the company. Because the first thing that had to happen was that an idea—regarding the classical rules of hotelling—had to be overturned. Namely, that hotels might not even be necessary.

It seems terribly strange that there are no safe or certain parameters for working with creative ideas. It would be so much better, wouldn't it, if we could simply practice being creative and applying it like a math formula. But creative uncertainty is our strength, and it is what makes humans irreplaceable even in the future. There is no algorithm, analyst, or efficient computer that can ever fully explain how we are able to come up with a creative idea. Creativity cannot be reduced to rules. It is the thing that changes the rules.

Switching on the lightbulb

AND SO HOW do we think of a new idea? At the very least, not by "producing" some concrete technique. Some companies have employed entire innovation departments and idea offices to come up with creative approaches and products on-site. But this is only the second-best approach. Our brain does not have an "idea department" that can be activated by a given method.

Because ideas don't get produced in a particular location of the brain, and they don't really get produced in this way in companies either. New ideas arise by breaking apart a problem, setting up flexible teams and different networks to search for a solution, and allowing them the space to work freely. But, even then, these methods don't generate finished ideas but rather idea bids that have to be tried out as quickly as possible, tested, and refined. Thus, finished ideas cannot be planned. They come about as a result of a creative process. This parallels the internal process in the brain, which is a back-and-forth interplay between distraction and concentration.

Present at the start of every good idea is pain—figuratively speaking. There is a disturbing problem that drives us to seek a solution. It is only when something really gets under our skin that we tend to do something about it. Irritation, not satisfaction, engenders creativity. We look for solutions when we are hungry, not when we are satiated. Happy people are not necessarily the ones changing the world. Because when we are happy, there is no need for change. To be optimistically dissatisfied is the right mode for creativity. In this state, we concentrate initially on the problem, and then we give ourselves space away to gather ideas, before finally narrowing them down in order to try out the best one. It is an idea loop.

There are all kinds of studies that show what can affect our creative abilities. For example, participants are more creative when there is a lightbulb hanging from the ceiling rather than a fluorescent tube.[14] But no one has (yet) studied whether or not the European Union is suffering innovation withdrawal from our ban on lightbulbs. Another effective suggestion is that movement helps our creativity. Participants who took a Torrance Test while walking on a treadmill had an 81 percent increase in creativity compared to the sedentary control group.[15] Even merely walking in place increases mental wandering. Like in one crazy mental experiment where participants are told to fantasize about totally unlikely things. For example, that one is nine feet tall, or can speak eight languages, or that the Florida Panthers will win the NHL Stanley Cup.[16] This kind of zany fantasizing results in more creative (or devious) ideas when participants take later tests.

Such experiments often highlight only tiny aspects of our creativity. They feed into our desire to somehow tame our unbridled creative thoughts. This is why there are countless techniques that we can follow to get our creative juices flowing: brainstorming, design thinking, the morphological matrix, the headstand technique, and many more. Don't get me wrong— all these techniques have definite benefits if used at the right time. But nonetheless they are all attempts at steering our seemingly unregulated creative power onto controllable tracks. Think back on the remote associates test, which could be solved either through analysis or insight. A creativity technique often suggests that creativity can be summoned by following a set of rules. But this is often only analytical thinking under a new guise. And though it may work, I doubt that the Otto engine, the bicycle, or the Yo app were generated in a coordinated brainstorming session like this.

Shaking up the flow of ideas

MANY CHAPTERS IN this book point out that creative ideas benefit from the many different characteristics and weaknesses of our brain. This enables us to combine concepts rapidly, which leads to new combinations of ideas, if also to false memories. Our indulgence in daydreams allows us to assume new perspectives. Distraction is an important source of inspiration, especially since creative people are very easily pulled out of concentration by new stimuli.

The power of clever and deliberate distraction in our default mode network should not be underestimated. Distraction is an important step that leads us to new perspectives and ideas. New perspectives are fostered by social contact with others and feedback is vital to developing our ideas. That's why we cannot Google or WhatsApp an idea. We have to be in personal contact with other people. This was precisely the problem in 2007 when a German bank could not get their new marketing campaign to work at all. When the bank analyzed the communication channels, they saw that all of their departments, from management to the developers to the customer service department, communicated well with one another. However, an assessment of the bank's modes of communication revealed that most information was transmitted via technology and not through personal (analog) conversations. The customer service department was located in a different part of the building, and so the employees did not participate in casual chats with those from other departments during breaks or in the hallways. It was only after the physical layout of departments was changed, and the relevant departments were moved closer to one another, that the new campaign was finally able to get off the ground.[17]

Often, it is the lack of personal interactions with others holding different perspectives that deprives us of valuable creative input (inspiration). One U.S.-based software company wanted to encourage such exchanges and so organized regular "beer meetings" and other social events in the hopes that the relaxed environment would help their employees to be fruitful and multiply (in terms of ideas). But an analysis of the communication behavior proved these meetings were insignificant. The creative flow of ideas improved, however, when the company then extended the size of its cafeteria dining tables. Suddenly employees were sitting next to unfamiliar coworkers.[18] Good ideas always arise when one is inspired by different perspectives and then pursues these ideas in a focused manner. Just like the balance between the default mode network and the control network: distracted idea gathering first, focused assessment second.

There is no right or wrong—there is only create

DO YOU REMEMBER the creativity puzzle from the beginning of the chapter in which you were asked to illustrate a word? This wasn't something I made up. It was a task given to participants in 2015 in order to study under what conditions people were especially creative when illustrating words such as "exhausted," "exactly," or "cry." People drew the most creative illustrations not so much when they deliberately considered the idea but when they activated their autopilot mode and simply started drawing without thinking too much about it.[19] In other words, when our conscious cerebrum lets go of the illustrative work, it is no longer monopolizing the situation and can instead give more space to creative ideas. This might be a reason why monotonous tasks are like creativity generators: driving a car, taking a shower, vacuuming. Ideas dawn on us much more readily whenever we are not dominated by our conscious thoughts. But this only works if we have been working intensively on a problem just prior to doing the mundane activity.

Using your knowledge of how the brain works, you can now develop your own "creativity technique." Regardless of what you choose to do, make sure to always toggle between activating your daydreaming default mode network and your sorting control network. Many authors, artists, and scientists (from Thomas Mann to Immanuel Kant to Beethoven) automatically followed this principle: they would first wrestle with a substantial issue for a period of time (sometimes for hours) before taking a break. Not a chaotic "I'm-going-to-start-something-new" kind of break but rather one in which they sought deliberate distance from the problem. Nowadays, we would call it chilling out by taking a walk, exercising, or doing chores around the house. The important thing is to look at breaks as a part of your work, since our brain is combining ideas into possible new solutions during a supposed period of "doing nothing." The potential ideas can then be harvested once you've returned to your work. Also important is that you set yourself regular and consistent time frames. Because we only trust new ideas when our condition and surroundings are safe and reliable. If we keep looking at the clock, getting sucked into our smartphones, or don't really have an idea how long a break should last, we won't be able to be creative because we will be inundated with new problems to solve.

There is no such thing as absolute creativity and no such thing as the perfect way to achieve it. There is only your way. In the same way that no one else could ever think your exact thoughts, only you are able to arrive at your very own kind of creativity. It cannot be calculated—whether your idea is good or bad in the end. This is precisely what makes your creativity unique and irreplaceable. Neither other people nor a computer can determine this for you.

So, what is the next big idea that will change the world? We don't have a clue. But we can be fairly certain that it will originate from a brain. Not because we are faster, more efficient, or more intelligent than machines but because we are the opposite: slow, imprecise, and prone to error. However, these reasons are also why we can understand things instead of merely analyzing them. We are able to consider new perspectives, and this gives us our unique mental strengths that we should put to use and be proud of. Because these strengths are what make us human.

14

PERFECTIONISM

Why We Need Mistakes in Order to Improve

T**HE DATE IS** June 24, 2010. At 4:47 p.m., one of the most historic moments in modern sports is about to end. John Isner stretches his fist up into the air at Wimbledon. After an epic eleven hours and five minutes, spread across three days, the longest tennis match in history is finally over. With a 6–4, 3–6, 6–7, 7–6, and 70–68 score, the U.S. tennis player wins against Nicolas Mahut. The fifth set alone lasted eight hours, setting the record for longest tennis match in history. The line umpires worked two-man shifts in order to maintain their concentration. Even the scoring technology reached its limits. The digital score cards stopped working at 47–47 because IBM had never programmed higher scores.[1] The reason behind this historically long game was that the players hardly made any mistakes. Prior to the decisive final point, both players had hit their serve eighty-four times without a single service break for over ten hours. John Isner especially demonstrated his precision in serving. At no time before or after has a player delivered so many service aces. He hammered 113 of them into his opponent's court. In his next match, John Isner lost without much ado in less than an hour and a half without hitting a single ace.

Isner had just previously set the world record for eternity and then was suddenly unable to ace a single serve. One certainly cannot accuse John Isner of not knowing how it's done. He had perfected the move and yet somehow still messed up. Why is it that he was suddenly no longer able to repeat what he had just achieved to perfection? "That's totally normal," you might say. "And only human. Mistakes are a part of life." That may be, but what is so good about being human? After all, inaccuracies and deviations are continuously seeping into our behaviors, and these make us more prone to error and inefficient. It's not a good thing, or is it?

Colloquially, we call these and other kinds of slipups "careless mistakes." There you are concentrating on something, when—oops!—you stop paying attention for a brief moment and the offense happens. An erroneous comma is placed in a sentence, the morning's cup of coffee is rattled and spilled, or your car throttle gets choked. Thousands of well-rehearsed and seemingly perfectly mastered routines veer off course. And while we were able to argue that there were hidden benefits behind the other weaknesses described in this book, this is, unfortunately, not true of superfluous blunders like these. Or, is there anything particularly positive about making a typo, getting someone's name wrong, or having to stick your foot in your mouth?

Given this, is it really any surprise that we try hard to avoid these kinds of mistakes? Perfection is our goal. Mistakes are for losers. But our brain is, unfortunately, a useless organ when it comes to achieving this goal because deviation and imprecision are a part of our neural network's method. Our brain naturally tries to process information as efficiently as possible. But at the same time, it also has to maintain some form of adaptability. We need adaptability in order to react if our surroundings change. Perfection in thinking is all well and good, but it is just as competitive as a monoculture cornfield: efficient and productive whenever conditions remain the same but quickly destroyed as soon as underlying conditions change.

Avoiding mistakes can also be a good thing but remember that just because you don't make a mistake does not necessarily imply that you are right. Even the most annoying careless mistakes and mental slips in our everyday lives serve to show us a very clever rule of the brain: what is important isn't avoiding mistakes but learning from them. Fortunately, our brains are masters at this.

Mistakes at the push of a button

CREATING MISTAKES AND then studying what is happening in the brain is child's play. To do so, you only need two ingredients: a task and one test person. That's it. If you are patient enough to wait, you can be pretty certain that sooner or later you will be able to harvest a mistake and study the corresponding brain functions. The tasks don't have to be all that complicated. On the contrary, we get into a mental lurch even with the simplest of problems if we have to repeat it often enough.

For example, cross out all of the lowercase e's on the previous page within thirty seconds. Or cross out only the e's found in words containing the letter n! Wait! Before you flip back a page and destroy this book for future readers, let's take a closer look at how we may be able to do this more systematically. For research, test conditions have to be standardized and one well-known concentration test that measures erroneous behavior is the d2 Test, which was created in the 1960s to test for the attention of car or truck drivers.

The d2 Test is another test in which a letter has to be crossed out, the letter d, whenever it is surrounded by two lines, or in this case, apostrophes. If there are more or fewer apostrophes around the d or a p instead of a d, you must not cross it out. For example, you should cross out

d   or   d

but not

d   or   p

Simple, right? But when you are looking at an entire list of d's and p's (and only have a half second per character), it is much harder:

d d p d d d p p d p d d d d d p d p d d d

This d2 attention test is comprised of a total of 658 charac-ters (fourteen rows with 47 characters each), of which 299 d2 characters have to be crossed out in only five minutes.[2] I'll spare you the entire cluttered list of letters and characters because we already know that careless mistakes are certainly going to be made. Test takers err on average almost ten times over the course of the test, by crossing out the wrong character or skip-ping over another. This may not seem like a big deal, but just consider how simple the task really is. The test becomes diffi-cult only when several of the simple tasks are piled onto each other. In addition, test takers get distracted by similar-looking characters and, when they are in a hurry, they easily overlook miniscule differences.

Disruptive symbols are also used in other laboratory tests—for example, the much-beloved-by-neuropsychologists Eriksen Flanker Task (named after both of the developers of the test). Similar to the above problem, the test is actually quite simple. Whenever a certain character appears in the middle position, participants are supposed to press on the right button and if the character does not appear, they should press on the left button. The designated character could be a letter, character, or an object. For example, if an M appears in the middle, you would press on the right button and if there is an N, press on the left. Next, rows of successive letters are displayed, and the test taker tries to react as quickly as possible to the target character:

MMMMM   MMNMM   NNMNN   NNNNN

Careless mistakes happen here as well. We suddenly let a wrong action slip out in the same way that a wrong word slips out when we misspeak. The more we are distracted by various stimuli, the more this happens. Such disruptions seem to roll out the red carpet to invite mistakes in. This is why tongue twisters work. You can easily say "seashells seashells seashells." But when you have to say "she sells seashells," you start to trip up, have to speak more slowly, or concentrate on your pronunciation. There's a good reason behind this. Because our brain cannot avoid such mistakes and, in fact, these mistakes are actually the result of the way in which our basic thinking functions.

The mistake competition at the center of the brain

WE OFTEN IMAGINE that an action occurs in our brain by a step-by-step process. And if we make a mistake, this must imply that something went wrong along the way. It's like a row of dominoes: one domino falls and pushes the next domino over until the entire pattern has fallen. Or if it doesn't happen, and one domino doesn't fall because it has been placed wrongly so that the row that follows it also does not fall. This is what we experience all around us: results occur via a step-by-step process. A then B. If I trip over a glass bottle, it falls over and breaks. If a mistake occurs, something went wrong somewhere.

But it is different in our brain. Yes, our actions and thought processes do follow steps. But in general, our thoughts do not follow a linear A-then-B logic. Instead, we generate a dynamic pattern of activities within our neural network—a thought murmur—from which a dominant directive eventually crystallizes. Once one of the many directives goes above a certain threshold, the other patterns give up, and only this particular directive is then followed.

To apply this tangibly to a mistake: if we are told to choose a d2 character from a long list of various characters, we have to first have a clear goal—for example, crossing out any letter d's that are surrounded or touched by two apostrophes. This plan is carried out by our prefrontal cortex. At the same time, we are constantly being bombarded with sensory stimuli that are recognized and processed in our brain's visual center (especially in the neck region). The factual difference in the visual center between a d and a p is already very clear. However, the visual center cannot plan an action; it can only provide the information, which the prefrontal cortex must carry out. To make things practical, halfway between the prefrontal cortex and the visual center is a region that reconciles both sides. This intermediary post is somewhat awkwardly called the basal ganglia.

Ganglia are something like nerve centers or relay stations that serve as an important junction for nerve connections. The basal (meaning basic) ganglia connect the neuron networks responsible for our movement and actions. They are located almost at the center of the brain, surrounded by the limbic system. This is where the all-important competition takes place for whether we make a mistake or not.

The talk-show principle

LET'S IMAGINE THAT we see the following in a d2 Test:

"
d

Right now, different patterns of action are competing in the circuits of our basal ganglia. There is the pattern that says, "Of course! Cross it out! It is a d2 character!" but there is also the pattern crying, "No, wait a sec, it looks just like the character I saw just a few seconds ago that wasn't a d2 letter!" or "It doesn't matter, the time is flying so just keep going and don't cross anything out!" Some of these patterns lead to a concrete action, some of them do not, but which one is fulfilled in the end depends on the specific goal as well as on the incoming sensory stimuli. The stronger the goal ("Choose the d2 character"), the sooner unsuitable patterns will be suppressed. Conversely, a whole lot of stimuli increase the likelihood that the wrong pattern will be executed. Within the neural network, there is thus a constant chaos of competing patterns that rise or fall dynamically until one of the patterns finally becomes so dominant that it sends itself as a directive to the movement centers of the brain. The winner takes it all—regardless of all the other patterns—only one is able to be 100 percent fulfilled.[3]

Imagine you are watching a political news show with representatives from various sides of an issue arguing. The discussion becomes heated, one side's facts are not the other side's facts, truth is not truth, and suddenly both people are talking over each other, and you cannot understand anything of what is being said. This is what happens in the basal ganglia. All of the patterns are running amok over each other with no clear result. The same way that the news viewers cannot understand what is being said, the brain also cannot carry out any task through the din of all of the different options. In the case of a news talk show, there are two possibilities: one person might begin speaking louder, drowning the other out until they finally fall silent, and the loudest opinion triumphs. But this is only the second-best option because the loudest mouth is often the most false. The second possibility is that a moderator, say CNN's Don Lemon or Erin Burnett, will break in and decide who gets to say something and when. When one person is speaking, the others must remain silent, but the moderator will often challenge the speaker to answer the question concretely. Such answers are often much more informative than the first possibility, but this method, unfortunately, does not happen enough on televised political discussions.

Let's apply this illustration to the brain: the talk-show program is the basal ganglia; the different presenters are the various patterns of action. And Erin Burnett is the cingulate cortex. The cingulate cortex is like the moderator in a talk show that plays a crucial role in singling out and later assessing an action. Just as a moderator judges whether a panelist's response is meaningful or nonsense, the cingulate cortex, as part of the prefrontal cortex, registers whether an action is correct or false. Without the cingulate cortex, we would not be able to quickly recognize mistakes or react to them accordingly.

The moderator must jump in

WE CAN ALREADY start to see that mistakes are generated in the brain when a "false" pattern is implemented over a better one. These dynamic, constantly tumultuous activities in the neural network, which grow increasingly stronger or weaker, are not as linear as a row of dominoes. Small changes in stimuli might help to strengthen a previously weaker pattern so that it suddenly becomes the dominant voice in the chaos. And then the mistake gets made.

Mistakes are inevitable because this system of action never functions perfectly. Certainly, the neurons adapt their connections over the course of time so that they can create an increasingly improved (and more frequent) pattern of action. But even a newly trained synapse that just performed very usefully might, at the next step, become impractical again and lead to a mistake. One can never know what the next challenge might bring. In the context of such a talking-over-one-another system, the brain is perfect at being imperfect because it is constantly employing adaptive actions. We are not programmed to fall like a line of dominoes. If we were, we would not be able to adapt.

Because we cannot avoid mistakes, we must learn to make the best of them. And to do this means to learn from them. Our brain has its mistake radar turned on high and is eager to alter its behavior every time we err. On a news-show panel, moderator Erin Burnett will dig deeper whenever a participant gives an answer that is completely off track. A brief but firm, "That's not answering my question!" gives the guest the opportunity to respond productively. Similarly, our brain issues a caution as well whenever there is a mistake: this is the ERN signal.

ERN stands for error-related negativity. If a participant taking the Eriksen Flanker Task (MMMMM or NNMNN?) makes a mistake, brain wave measurements show a reduction in voltage (thus the negativity). This activity is generated only a tenth of a second after the mistake has been made.[4] This is so fast that one has not yet even consciously registered that one has made a mistake. But the brain has registered the mistake, and this has an effect on our behavior.

The brain that cried "Error!"

WHENEVER WE MAKE a mistake, there are usually two things that happen: Firstly, we hesitate briefly and thereby start operating at a slower pace. At the same time, we start to pay more attention to our actions, which means that we seldom make the same mistake two times in a row. Both effects appear to be mediated through the ERN signal in our brain. Research has shown that following a heavy deceleration, study participants generate a remarkably strong ERN signal[5] and that the ERN signal is strengthened whenever they correct their mistakes especially rapidly and with careful attention.[6]

This cry of "Mistake!" apparently helps the brain to better filter out the various action patterns, before one is able to dominate over the others and subsequently be carried out. If there were no moderator, talk-show panelists would continue to talk over one another until one of them starts to shout. There is one advantage to this mode of discussion. There are so many different arguments, and the discussion could go in any direction. But it is only when the loudest voices are quieted so that the softer voices may have a say that previously hidden, and also potentially meaningful responses, have a chance to rise to the surface. This response filter also takes place in the brain, as the ERN signal more or less depicts the influence of the moderating cingulate cortex. The stronger the signal, the more the impulsive and pushy action patterns (which are usually false) are filtered out.[7] At the same time, the movement regions calm down somewhat, as though they were saying, "Wait a minute! Before we decide on a course of action, the basal ganglia need a moment to sort all the patterns out." This brief delay can be helpful in sorting out the better patterns so that the only patterns that remain are those which suit the task. In other words: those that are correct. This is the reason why we tend to be slower just following a mistake, but therefore we are also more precise.

The benefit of chaos

IT ONLY TAKES one careless mistake to highlight how well-balanced the brain is when planning actions and carrying them out. Indeed, it has to fulfill two contradictory tasks. On the one hand, it must work precisely in order to avoid dumb and impulsive mistakes. And on the other, it must be adaptive and diligent at testing enough different options to be able to respond to a changing environment. That is, efficiently recognizing, eradicating, and avoiding mistakes while at the same time inefficiently thinking up new actions (as described in the previous chapter on creativity).

For this reason, the brain uses the best of both worlds: it risks making a mistake every now and then because it generates numerous competing patterns, one of which might occasionally be false. For example, if a distracting signal pushes onto the wrong track, the sensory-stimulated patterns are dominant, and we make a mistake (just think of "she sells seashells" or other tongue twisters). But the brain is quick to recognize these mistakes and adapts its behavior. We are constantly gathering feedback in order to improve.

If our brains were to switch to a logical system of thought, we would lose our mental flexibility entirely. A careless mistake doesn't usually do too much damage. Whereas it would be boring if we were only ever logical, calculated, and free of even small errors. After all, we don't live in a static world where a perfectly efficient action leads to success every time. Change is what propels us forward. And sometimes it is better to go off course from our original plans.

How the computer came to master chess

MAY 1997. THE supremacy of the human brain is at stake. Garry Kasparov, the reigning world chess champion, gets up from the table in irritation and concedes the chess game. It has now become clear that, for the first time, a chess program developed by IBM, called Deep Blue, has won a six-game match against a chess world master. It's a sensation! The computer is now mentally superior to humans!

In those days, the chess computer was as large as a closet, a milestone in computer development. These days, a chess program has been reduced to the size of an app that can run on your phone for a couple of dollars and is so good that it will never lose a game against a human. Deep Blue was able to calculate 200 million positions per second. Current-day programs don't even calculate 10 million positions but are, nonetheless, so efficient that they focus only on the most important moves and thus never err. That's why no chess world master has ever won a game against the best and most updated chess program since 1997.

The decisive game for Deep Blue's victory was the first one of six.[8] Kasparov was winning but in the 44th move, the machine chose a very strange variant, a move that Kasparov had never before seen. It was not a dumb move and, after a few further moves, it had the potential to seem as if it had been a clever move. Kasparov was baffled. No logically thinking human would ever think of a move like that. Was this perhaps the first flicker of real artificial creativity? Unthinkable, though possible. During the second game, the computer continued its mysterious game. Deep Blue once again made a strange move that contradicted everything that was previously understood about chess strategy. Kasparov, a master in psychological warfare, believed that there must be other expert chess players prompting the computer from behind the scenes because such an undigital form of creativity was inexplicable for a computer. Kasparov was so outraged over this that he got flustered and lost the second game. He was unable to recover from this decisive blow in later games and, in exasperation, finally lost the entire tournament. Afterward, when programmers studied the machine to find out why it had chosen such unusual and inexplicable moves, they realized that the chess program had become overloaded at certain moments. In the first game, it chose a random move to avoid crashing and for the purpose of simply offering any kind of move. A classical computer error that nonetheless led to the machine winning the duel. It did not win because it was better at calculating (over the course of the subsequent games, the computer demonstrated several weaknesses, showing that Kasparov was in fact the superior player), but because it made a supposed error at the right moment.

Perfectly boring

THERE ARE THREE things that stand out on this footnote of computer history. Firstly, error-free behavior does not always lead to success. Secondly, this was probably the first, only, and last moment in the history of computers that a machine was really creative. Because it is only through a small and unexpected mistake that it is possible to break free of a programmed and fixed structure. Thirdly, this example is also a good model for how we should not deal with mistakes. Because in the end, IBM dismantled Deep Blue, snuffed out the project, and did not continue to pursue the idea that even computers can achieve success through illogical errors.

Modern chess programs are unbeatable—not because they reinterpret unusual chess moves and thus upstage their opponents—but because they simply wait until the human makes a mistake. As this is certain to happen. The program does not win the prize for most congenial chess player, but the prize for most perfectionist player. The same is true of algorithms that play poker or the Asian board game go. The programs play the various games several (hundred thousand) times against themselves and adapt their technique so they will never lose. But at the end of the day, these programs are not creative masters in their field. They are only extraordinarily good defense artists. They don't beat their opponents so much as just wait until their opponents hurt themselves. It's a bit like a soccer team that holds back on the defensive line and bides its time until the other team misses the penalty kick. If you are playing against a human, this is a pretty good strategy. But it is also a boring one. If two perfect chess programs are squared off against each other, the game will always end with a draw. Error-free decision-making by two opponents leads to a complete balance of power (in mathematics this is referred to as the Nash equilibrium). The tennis match between Isner and Mahut was only so exciting and spectacular because onlookers knew that it would eventually come to an end. By the way, Isner won the match with a first-class passing shot, a real winner. He defeated Mahut and did not anticipate his opponent's mistake. If the game had involved two perfect tennis robots playing against each other, the match would still be going on today. But no one would be watching.

An error-free world is not necessarily a progressive world. On the contrary: it would be static, stable, and hostile to advancement. Without the risk that one might err, there can also be no courage for trying something new. New ideas are not formed by thinking that is exclusively intelligent (fast and error-free), but only when we allow ourselves the possibility of messing up every now and then. All of the futuristic scenarios in which superintelligent computers assume world domination are thus not quite accurate. World domination requires more than intelligence. You also have to be a bit crazy, willing to break the rules and standards instead of always conforming to them—and in order to do these things, you have to risk making a mistake.

Systematic errors

OUR BRAIN HAS largely systematized mistakes. It is not set on thinking perfectly right off the bat but allows for occasional lapses. This can be successful or disastrous; it's impossible to know beforehand. There are many scientific, cultural, or economic advances that owe their origins to more or less chance errors. Alexander Fleming erroneously let his petri dish develop mold—and discovered penicillin. Édouard Bénédictus bumped a glass chemistry flask that had not been properly cleaned out. When it fell to the floor and cracked, he found that his original mistake had caused a film of plastic to coat the inside of the flask—thus safety glass was born. The engineer Percy Spencer was standing in front of a giant electromagnetic radiation device when he noticed that the chocolate bar in his pocket had melted—he thereafter developed the microwave. Making chance mistakes is one thing. Using them is another. Or, as Louis Pasteur said, "Fortune favors the prepared mind."

Don't misunderstand me. I am not arguing that we should let our groceries go bad or forget to clean out our beakers. And while we're on it, there are lots of other reasons that chocolate bars can melt in your trouser pockets. What I would like to emphasize is that it is our exaggerated longing for flawlessness that freezes our thinking and makes it blind to the potential uses of some errors. As soon as our brain makes a mistake, it not only tries to correct it, it also productively uses it. It is precisely because a mistake offers the potential for improvement that this flawed system of thinking has prevailed in evolution. This exposes the brain to the risk that it might generate something awful. But this is a price well paid for our ability to remain flexible. If we, instead, only functioned according to an efficient, error-free thought system, we would have been eradicated with the first serious environmental change.

The art is not in avoiding mistakes. Anyone who tries this will eventually become as boring as a chess computer. And, what's worse, replaceable. Because sooner or later algorithms will be able to avoid mistakes and carry out an action efficiently and flawlessly. But recognizing whether a supposed mistake might yet have a purpose is an ability that only humans can claim.

Of course, a mistake made in the d2 attention test is never good, creative, or productive. Nevertheless, concentration tests have helped us enormously in brain research. Now we understand that the brain almost systematically incorporates errors in order to examine them and to alter its behavior. And this is the important lesson that should be drawn from such tests: to err is human—and, for the brain, is extremely useful.

Better done than perfect

WHAT CAN BE learned by making mistakes? The simplest careless mistakes in concentration tests show what the brain does with mistakes. Firstly, it makes them and does not try at all costs to avoid them. Secondly, when it has erred, it hesitates briefly and then adapts its behavior to prevent, as much as possible, repeating the same mistake (test takers, for example, slow down and concentrate more). Thirdly, the brain keeps right on going. It does not alter its thinking strategy. And for this reason, it will continue to make careless mistakes. Perhaps not in exactly the same context as before—but in a different one. But it is inevitable, and this is why a mistake is more than a mental failure. It is rather a signal to be a little bit better.

Attention tests (such as the d2) show that it is not always a lack of concentration that leads us to make a careless mistake. If that were true, mistakes would occur more and more as the test progressed and would be randomly distributed. But analysis of a few hundred test takers shows the opposite. The mistakes don't occur randomly because a test taker here or there has an attention lapse. Instead, the mistakes made by numerous test takers are clustered around particular regions of the test itself, in completely inconspicuous spots, as though the characters assembled there especially prompted errors.[9] The reason for this might be that false activity patterns in the brain are favored and then put into action in the basal ganglia. The brain is always trying out new activities and sometimes even chooses the wrong ones to set in motion. But the brain nonetheless risks it.

Trying things out and taking into account that mistakes will be made is a terrific strategy for arriving at new knowledge. Because if you are interested in learning something new, an initial mistake is a good sign. It shows that you have tried (even if it didn't go too well). But now your experience has given you a much more direct access to a problem than if you had only theoretically thought through the entire scenario. This is a significant advantage on the way to understanding correlations. Every scientist could sing a tune on this topic. At least, I've never met anyone who did not understand their own research through trying things out and making mistakes. You can't discover new knowledge in books, after all. You have to generate it. And this is only possible when we risk being wrong.

Even when the knowledge already exists and "merely" has to be conveyed, the principle of making mistakes and trying out variations still applies. Because whether one is in school or working, there are basically only two ways to convey knowledge. Firstly, you could explain a fundamental concept and then apply it practically. You could explain, for example, the mathematical principle of standard deviation and then follow up with a few exercises to practice it. But this is only the second-best option. In studying the learning abilities of ninth graders, scientists have found that it is better for students to first try out actual examples, to fool around with different solutions, and even sometimes to fail. After this, students are much more receptive to the underlying mathematical concept and do not forget it as easily once it has been explained to them.[10]

Learning from mistakes

RISKING MISTAKES IS only the first step. Of equal importance is receiving feedback for your actions and then adapting your behavior. In the aforementioned concentration tests, this happens within fractions of a second. In the context of other learning processes, it may take a bit longer, but the same basic principle applies.

The more substantial the learning process is, the more time should be taken before feedback is given. One might assume that it's better to receive a reward or punishment directly after each mistake, as if we were raising a dog. If the dog behaves well, we toss it a treat. But if it does anything wrong, the dog is punished. However, this is training, not learning new knowledge. It is much better to wait for a period of time before offering feedback. When test participants are told to read a text (for example, about the sun) and then answer a multiple-choice test about their newly acquired knowledge, it makes a difference whether they are apprised of the correct answer immediately after each of their responses, or whether they are only informed about the right answers ten minutes later. Those who were given a short break in between were able to retain their knowledge much better and could still recall it on the next day.[11] You might say we are mentally disposed for delayed feedback, which is much more effective in the long run.

It is very important that the critique of one's mistakes is not offered as a personal failure. Even geniuses make mistakes— this might be one of the reasons why they are geniuses in the first place. By delaying feedback somewhat, we are able to avoid immediate punishment, which can easily be taken personally. It is absolutely crucial to avoid an environment of fear.

This is already evident in the development of small children. A parenting style that is perfectionistic ruins the self-esteem of three- to twelve-year-old children. Specifically, researchers have studied the ways in which parental pressure for children to be free of mistakes contributes to their children's anxiety. Take note: the more the parents concentrated on eradicating every one of their child's mistakes, the more distraught their smaller ones became.[12] Studies of the rhetoric used by parents showed that the anxious children were often spoken to with negative terms ("You are doing that wrong!"). This serves to undermine the brain's use of its mistake system in children. Instead of acting like a know-it-all when a child makes a mistake ("It's not I *goed* to the library, it's I *went*! W-E-N-T!"), it would be better to give a good example, ignore the mistake, and exemplify the correct way ("Oh really, you *went* to the library?").

One's environment thus plays a decisive role in whether a mistake is used to encourage new ideas or whether it is used as social ostracism. Of course, there are areas in which mistakes have to be avoided. For example, one should take care that nothing goes wrong when producing food, landing an airplane, or installing an electrical cable. If you drive a car, you should follow the traffic rules, absolutely. But if you're going to invent a new car or future means of transportation, you have to change the rules and break your current thinking patterns. Many things around us in our daily or business life should all naturally happen as efficiently and flawlessly as possible. But almost all of these tasks can principally also be automated, in contrast to the true achievement of our brain—namely, to acquire new knowledge. This only works, however, if we dare to make a mistake.

Onward, always onward!

REGARDLESS OF WHETHER and how one has made mistakes, the brain keeps on marching without changing its fundamental system of thought that allows for errors to happen. Of course, it switches on its filter mechanisms to avoid making the same mistakes all over again next time. But its fundamental operational principle is never shaken. It is not afraid of tripping over the next mental pitfall. For good reason.

When studying how the brain functions when making and correcting mistakes, one notes that several extended networks are active (for sensory processing, motion planning, the basal ganglia, the planning and assessing prefrontal cortex), but there is one important region often missing: the one responsible for fear. As bad as mistakes can be, we do not seem to possess an in-built fear of them. This implies that the brain does not punish itself for an error. It is only when we are socialized that a mistake is something bad that we develop a fear of doing something wrong. This is exactly what our adaptable brain doesn't need. Because whoever is afraid of making a bad decision will never find the right one. Or, what's worse, will be too afraid to ever act at all.

There is nothing particularly terrifying about an error per se. And this is how it should be. Because new ways of thinking often only arise through making mistakes. Okay, maybe not when you are taking a concentration test—but real life is much more complex than simply crossing out a bunch of letters in a sequence. At the moment, my young neighbor is learning to ride a bike. I can guarantee you that he is going to crash at some point (and as an avid road biker myself, I know what that looks like). But that's the definition of riding a bike: falling over and getting back into the saddle again. My neighbor is certain to get scuffed-up knees, but in the best case he will also go places on his bicycle that he never knew before.

What this also requires is a certain amount of willingness to let mistakes happen without punishment. Not an easy task in our society, where mistakes have a negative image. Errors, defeat, falling flat on our faces—in some places these terms mean you're an idiot. In others, like Silicon Valley, they might mean you're on your way to the next big investment. After all, a tech developer who has never totally botched a start-up or two will probably also be unattractive to potential investors, who might be thinking, "Well, they are bound to make a mistake at some point, and I don't want them to do it on my dime." But someone who has already failed can now prove that they can recover from it. This is the only way to start up the next high-tech company. Or, how about these innovative new ideas for the amazing Californian market leaders: a company that permanently improves the potholes on Californian streets, or solves the problem of traffic on the Bay Bridge, or else finally invents doors in the U.S. that shut tightly and keep out the draft? A little bit of German precision could go a long way in Silicon Valley.

Errors are best put to use when we can make them in an anxiety-free environment. In the brain, mistakes are not an impetus for punishment but a good opportunity to reassess one's own thinking. Progress is only possible when we dare to make a mistake.

Stay fallible

THIS BOOK HAS taken you through many of the brain's weaknesses and errors. Some of them are really annoying and dumb, like blackouts in front of an audience, our weakness for distracting smartphones, unnecessary, careless blunders. Other flaws merely mask the hidden strengths of our brain—for example, that we create false memories, are not very good at numbers, or don't have a good feeling for the time. Regardless of the type of error, all of them come about because our brain does not care about being flawless and perfect. Because if it did, it would not be very flexible.

Careless mistakes are a prime example of how the brain factors in mistakes. It employs an operational system that is not logical and linear but seems to be helter-skelter and disorderly. Because of this, mistakes are inevitable, but this is not a bad thing because our brain is not afraid of doing something wrong. If we never erred, we would never be able to change. We would then not only be incapable of learning and boring as heck, we would also, sooner or later, be easily replaced by a more efficient computer.

Instead of kicking ourselves for every mistake, we should be happy that we are free enough to make them. And we shouldn't punish ourselves or others for mental lapses. The unique characteristic of human thought lies precisely in that it is not flawless and exact. Error-prone thinking is the only thing making us superior to machines and sets us apart from noncreative computers. Essentially, our weaknesses in thought are really our greatest mental secret weapons. Although we rightly should not jump for joy over every error, blunder, or aberration, it is much more important that we do not feel afraid of them. Because mistakes give us the cognitive edge.

So, remain as imperfect as you are. It's what makes you unique. Make mistakes and, in doing so, give rise to new ideas. After all, to err is what we do best.

NOTES

Y<small>OU CAN GOOGLE</small> data. If you are really good, you can Google information. But Googling knowledge is quite a bit harder. Because knowledge is what happens when we use information to change our ways of thinking. And this is going to remain an analog process for the foreseeable future.

The quality of thinking naturally rises and falls with whatever source of information one uses. Nowadays it is easier than ever to access information. But this is all the more reason to pay attention to the quality and relevance of information sources. The median date of the scientific publications listed here is therefore 2012. In addition, almost 90 percent of the material is from scientific publications in peer-reviewed journals—publications in which you cannot simply write whatever suits your fancy. Science thrives on critical discourse, hypotheses, failures, and experiments. Scientific truth is not something that can be decided by votes, but we are able to get closer to it inch by inch. Or, as my chemistry teacher said: "Whatever you do, whatever you think or research—nature is always right. It doesn't make any mistakes. To err is a uniquely human propensity."

Chapter 1: Forgetting

1. Blake AB et al. (2015) "The Apple of the mind's eye: Everyday attention, metamemory, and reconstructive memory for the Apple logo," *Q J Exp Psychol*, 68(5):858–65

2. Castel AD et al. (2012) "Fire drill: Inattentional blindness and amnesia for the location of fire extinguishers," *Atten Percept Psychophys*, 74(7):1391–6

3. Snyder KM et al. (2014) "What skilled typists don't know about the QWERTY keyboard," *Atten Percept Psychophys*, 76(1):162–71

4. Martin M, Jones GV (1998) "Generalizing everyday memory: Signs and handedness," *Mem Cognit*, 26(2):193–200

5. Wimber M, Alink A, Charest I, Kriegeskorte N, Anderson MC (2015) "Retrieval induces adaptive forgetting of competing memories via cortical pattern suppression," *Nat Neurosci*, 18(4):582–9

6. Dunsmoor JE et al. (2015) "Emotional learning selectively and retroactively strengthens memories for related events," *Nature*, 520(7547):345–8

7. Mosha N, Robertson EM (2016) "Unstable memories create a high-level representation that enables learning transfer," *Curr Biol*, 26(1):100–5

Chapter 2: Learning

1. Hermans EJ, Henckens MJ, Joëls M, Fernández G (2014) "Dynamic adaptation of large-scale brain networks in response to acute stressors," *Trends Neurosci*, 37(6):304–14

2. Strelzyk F, Hermes M, Naumann E, Oitzl M, Walter C, Busch HP, Richter S, Schächinger H (2012) "Tune it down to live it up? Rapid, nongenomic effects of cortisol on the human brain," *J Neurosci*, 32(2):616–25

3. Schwabe L, Wolf OT (2010) "Learning under stress impairs memory formation," *Neurobiol Learn Mem*, 93(2):183–8

4. McGaugh JL (2013) "Making lasting memories: Remembering the significant," *Proc Natl Acad Sci USA*, 110 Suppl 2:10402–7

5. Draschkow D, Wolfe JM, Võ ML (2014) "Seek and you shall remember: Scene semantics interact with visual search to build better memories," *J Vis*, 14(8):10, 1–18

6. Kornell N, Bjork RA (2008) "Learning concepts and categories: Is spacing the 'enemy of induction'?" *Psychol Sci*, 19(6):585–92

7.   Smolen P, Zhang Y, Byrne JH (2016) "The right time to learn: Mechanisms and optimization of spaced learning," *Nat Rev Neurosci*, 17(2):77–88

8.   www.tagesanzeiger.ch/digital/wild-wide-web/google-pixelt-kuhkoepfe/story/28203914

9.   Markson L, Bloom P (1997) "Evidence against a dedicated system for word learning in children," *Nature*, 385(6619):813–5

10. Childers JB, Tomasello M (2003) "Children extend both words and non-verbal actions to novel exemplars," *Developmental Science*, 6(2):185–90

11. Coutanche MN, Thompson-Schill SL (2015) "Rapid consolidation of new knowledge in adulthood via fast mapping," *Trends Cogn Sci*, 19(9):486–8

12. Nguyen A, Yosinski J, Clune J (2015) "Deep neural networks are easily fooled: High confidence predictions for unrecognizable images," *Computer Vision and Pattern Recognition* (CVPR '15), IEEE, 427–36

13. Thorne KJ, Andrews JJ, Nordstokke D (2013) "Relations among children's coping strategies and anxiety: The mediating role of coping efficacy," *J Gen Psychol*, 140(3):204–23

Chapter 3: Memory

1.   articles.latimes.com/1996-11-23/local/me-2006_1_tom-rutherford

2.   Wells GL, Memon A, Penrod SD (2006) "Eyewitness evidence: Improving its probative value," *Psychol Sci Public Interest*, 7(2):45–75

3.   Howe ML, Knott LM (2015) "The fallibility of memory in judicial processes: Lessons from the past and their modern consequences," *Memory*, 23(5):633–56

4.   Lacy JW, Stark CE (2015) "The neuroscience of memory: Implications for the courtroom," *Nat Rev Neurosci*, 14(9):649–58

5.   Stadler MA, Roediger HL, McDermott KB (1999) "Norms for word lists that create false memories," *Mem Cognit*, 27(3):494–500

6.   Kim H, Cabeza R (2007) "Differential contributions of prefrontal, medial temporal, and sensory-perceptual regions to true and false memory formation," *Cereb Cortex*, 17(9):2143–50

7.   Straube B, Green A, Chatterjee A, Kircher T (2011) "Encoding social interactions: The neural correlates of true and false memories," *J Cogn Neurosci*, 23(2):306–24

8.   Pardilla-Delgado E, Alger SE, Cunningham TJ, Kinealy B, Payne JD (2015) "Effects of post-encoding stress on performance in the DRM false memory paradigm," *Learn Mem*, 23(1):46–50

9.   Bland CE, Howe ML, Knott L (2016) "Discrete emotion-congruent false memories in the DRM paradigm," *Emotion*, 16(5):611–9

10.   Stark CE, Okado Y, Loftus EF (2010) "Imaging the reconstruction of true and false memories using sensory reactivation and the misinformation paradigms," *Learn Mem*, 17(10):485–8

11.   Hupbach A, Gomez R, Hardt O, Nadel L (2007) "Reconsolidation of episodic memories: A subtle reminder triggers integration of new information," *Learn Mem*, 14(1-2):47–53

12.   Edelson M, Sharot T, Dolan RJ, Dudai Y (2011) "Following the crowd: Brain substrates of long-term memory conformity," *Science*, 333(6038):108–11

13.   Otgaar H, Candel I, Merckelbach H, Wade KA (2008) "Abducted by a UFO: Prevalence information affects young children's false memories for an implausible event," *Applied Cognitive Psychology*, 23(1):115–25

14.   Shaw J, Porter S (2015) "Constructing rich false memories of committing crime," *Psychol Sci*, 26(3):291–301

15.   Dennis NA, Johnson CE, Peterson KM (2014) "Neural correlates underlying true and false associative memories," *Brain Cogn*, 88:65–72

16.   Carmichael AM, Gutchess AH (2016) "Using warnings to reduce categorical false memories in younger and older adults," *Memory*, 24(6):853–63

17.   Petersen N, Patihis L, Nielsen SE (2015) "Decreased susceptibility to false memories from misinformation in hormonal contraception users," *Memory*, 23(7):1029–38

18.   Bradfield AL, Wells GL, Olson EA (2002) "The damaging effect of confirming feedback on the relation between eyewitness certainty and identification accuracy, " *J Appl Psychol*, 87(1):112–20

19.   Josephs EL, Draschkow D, Wolfe JM, Võ ML (2016) "Gist in time: Scene semantics and structure enhance recall of searched objects," *Acta Psychol (Amst)*, 169:100–8

20.   Hunt K, Chittka L (2014) "False memory susceptibility is correlated with categorisation ability in humans," *F1000Res*, 3:154

21.   Howe ML, Wilkinson S, Garner SR, Ball LJ (2016) "On the adaptive function of children's and adults' false memories," *Memory*, 24(8): 1062–77

22. Schacter DL, Addis DR, Buckner RL (2007) "Remembering the past to imagine the future: The prospective brain," *Nat Rev Neurosci*, 8(9): 657–61

23. Wilson AE, Ross M (2001) "From chump to champ: People's appraisals of their earlier and present selves," *J Pers Soc Psychol*, 80(4):572–84

Chapter 4: Blackout

1. Barlowa M, Woodman T, Gorgulua R, Voyzey R (2016) "Ironic effects of performance are worse for neurotics," *Psychology of Sport and Exercise*, doi: 10.1016/j.psychsport.2015.12.005

2. Beilock SL, Bertenthal BI, McCoy AM, Carr TH (2004) "Haste does not always make waste: Expertise, direction of attention, and speed versus accuracy in performing sensorimotor skills," *Psychon Bull Rev*, 11(2):373–9

3. Beilock SL, Decaro MS (2007) "From poor performance to success under stress: Working memory, strategy selection, and mathematical problem solving under pressure," *J Exp Psychol Learn Mem Cogn*, 33(6):983–98

4. Lyons IM, Beilock SL (2012) "When math hurts: Math anxiety predicts pain network activation in anticipation of doing math," *PLoS One*, 7(10):e48076

5. Yoshie M, Kudo K, Ohtsuki T (2009) "Motor/autonomic stress responses in a competitive piano performance," *Ann N Y Acad Sci*, 1169:368–71

6. Yoshie M, Nagai Y, Critchley HD, Harrison NA (2016) "Why I tense up when you watch me: Inferior parietal cortex mediates an audience's influence on motor performance," *Sci Rep*, 6:19305

7. Mobbs D, Hassabis D, Seymour B, Marchant JL, Weiskopf N, Dolan RJ, Frith CD (2009) "Choking on the money: Reward-based performance decrements are associated with midbrain activity," *Psychol Sci*, 20(8):955–62

8. Autin F, Croizet JC (2012) "Improving working memory efficiency by reframing metacognitive interpretation of task difficulty," *J Exp Psychol Gen*, 141(4):610–8

9. Balk YA, Adriaanse MA, de Ridder DT, Evers C (2013) "Coping under pressure: Employing emotion regulation strategies to enhance performance under pressure," *J Sport Exerc Psychol*, 35(4):408–18

Chapter 5: Time

1. www.stiftungfuerzukunftsfragen.de/newsletter-forschung-aktuell/266/

2. Roy MM, Christenfeld NJ, McKenzie CR (2005) "Underestimating the duration of future events: Memory incorrectly used or memory bias?" *Psychol Bull*, 131(5):738–56

3. Buehler R, & Griffin D (2003) "Planning, personality, and prediction: The role of future focus in optimistic time predictions," *Organizational Behavior and Human Processes*, 92:80–90

4. Roy MM, Christenfeld NJ (2007) "Bias in memory predicts bias in estimation of future task duration," *Mem Cognit*, 35(3):557–64

5. Ogden RS (2013) "The effect of facial attractiveness on temporal perception," *Cogn Emot*, 27(7):1292–304

6. Effron DA, Niedenthal PM, Gil S, Droit-Volet S (2006) "Embodied temporal perception of emotion," *Emotion*, 6(1):1–9

7. Stetson C, Fiesta MP, Eagleman DM (2007) "Does time really slow down during a frightening event?" *PLoS One*, 2(12):e1295

8. Haggard P, Clark S, Kalogeras J (2002) "Voluntary action and conscious awareness," *Nat Neurosci*, 5(4):382–5

9. Stetson C, Cui X, Montague PR, Eagleman DM (2006) "Motor-sensory recalibration leads to an illusory reversal of action and sensation," *Neuron*, 51(5):651–9

10. Van der Burg E, Goodbourn PT (2015) "Rapid, generalized adaptation to asynchronous audiovisual speech," *Proc Biol Sci*, 282(1804):20143083

11. Wittmann M (2013) "The inner sense of time: How the brain creates a representation of duration," *Nat Rev Neurosci*, 14(3):217–23

12. Hancock PA, Rausch R (2010) "The effects of sex, age, and interval duration on the perception of time," *Acta Psychol* (Amst), 133(2):170–9

13. Zivotofsky AZ, Eldror E, Mandel R, Rosenbloom T (2012) "Misjudging their own steps: Why elderly people have trouble crossing the road," *Hum Factors*, 54(4):600–7

14. van de Ven N, van Rijswijk L, Roy MM (2011) "The return trip effect: Why the return trip often seems to take less time," *Psychon Bull Rev*, 18(5):827–32

15. Sackett AM, Meyvis T, Nelson LD, Converse BA, Sackett AL (2010) "You're having fun when time flies: The hedonic consequences of subjective time progression," *Psychol Sci*, 21(1):111–7

Chapter 6: Boredom

1. Deco G, Corbetta M (2011) "The dynamical balance of the brain at rest," *Neuroscientist*, 17(1):107–23

2. Leech R, Sharp DJ (2014) "The role of the posterior cingulate cortex in cognition and disease," *Brain*, 137(Pt 1):12–32

3. Utevsky AV, Smith DV, Huettel SA (2014) "Precuneus is a functional core of the default-mode network," *J Neurosci*, 34(3):932–40

4. Smallwood J, Schooler JW (2015) "The science of mind wandering: Empirically navigating the stream of consciousness," *Annu Rev Psychol*, 66:487–518

5. Wilson TD, Reinhard DA, Westgate EC, Gilbert DT, Ellerbeck N, Hahn C, Brown CL, Shaked A (2014) "Just think: The challenges of the disengaged mind," *Science*, 345(6192):75–77

6. Havermans RC, Vancleef L, Kalamatianos A, Nederkoorn C (2015) "Eating and inflicting pain out of boredom," *Appetite*, 85:52–57

7. Killingsworth MA, Gilbert DT (2010) "A wandering mind is an unhappy mind," *Science*, 330(6006):932

8. www.pressebox.de/inaktiv/avg-technologies-ger-gmbh/Weltweite-AVG-Umfrage-zeigt-SexOE-Nein-danke-Knapp-die-Haelfte-der-deutschen-Frauen-verzichtet-lieber-auf-Sex-als-auf-ihr-Smartphone/boxid/606696 (German only)

9. Britton A, Shipley MJ (2010) "Bored to death?" *Int J Epidemiol*, 39(2):370–1

10. Danckert J, Merrifield C (2016) "Boredom, sustained attention and the default mode network," *Exp Brain Res*, doi: 10.1007/s00221-016-4617-5

11. Smallwood J, Andrews-Hanna J (2013) "Not all minds that wander are lost: The importance of a balanced perspective on the mind-wandering state," *Front Psychol*, 4:441

12. Baird B, Smallwood J, Mrazek MD, Kam JW, Franklin MS, Schooler JW (2012) "Inspired by distraction: Mind wandering facilitates creative incubation," *Psychol Sci*, 23(10):1117–22

13. Hao N, Wu M, Runco MA, Pina J (2015) "More mind wandering, fewer original ideas: Be not distracted during creative idea generation," *Acta Psychol* (Amst), 161:110–6

14. Garrison KA, Zeffiro TA, Scheinost D, Constable RT, Brewer JA (2015) "Meditation leads to reduced default mode network activity beyond an active task," *Cogn Affect Behav Neurosci*, 15(3):712–20

Chapter 7: Distraction

1. www.careerbuilder.com/share/aboutus/pressreleasesdetail.aspx?sd=6/12/ 2014&id=pr827&ed=12/31/2014

2. www.commonsensemedia.org/sites/default/files/uploads/research/census_ executivesummary.pdf

3. www.symantec.com/content/dam/symantec/docs/reports/istr-21-2016-en.pdf

4. www.incapsula.com/blog/bot-traffic-report-2015.html

5. Lavie N, Tsal Y (1994) "Perceptual load as a major determinant of the locus of selection in visual attention," *Percept Psychophys*, 56(2):183–97

6. Gaspar JM, Christie GJ, Prime DJ, Jolicœur P, McDonald JJ (2016) "Inability to suppress salient distractors predicts low visual working memory capacity," *Proc Natl Acad Sci USA*, 113(13):3693–8

7. Feng S, D'Mello S, Graesser AC (2013) "Mind wandering while reading easy and difficult texts," *Psychon Bull Rev*, 20(3):586–92

8. Salomon R et al. (2016) "The insula mediates access to awareness of visual stimuli presented synchronously to the heartbeat," *J Neurosci*, 36(18):5115–27

9. Simons DJ, Chabris CF (1999) "Gorillas in our midst: Sustained inattentional blindness for dynamic events," *Perception*, 28(9)1059–74

10. Drew T, Võ ML, Wolfe JM (2013) "The invisible gorilla strikes again: Sustained inattentional blindness in expert observers," *Psychol Sci*, 24(9):1848–53

11. www.dekra.de/de/fussgaenger-beim-ueberqueren-der-strasse-riskante-ablenkung-durch-smartphones/

12. Rees G, Frith CD, Lavie N (1997) "Modulating irrelevant motion perception by varying attentional load in an unrelated task," *Science*, 278(5343):1616–9

13. Molloy K, Griffiths TD, Chait M, Lavie N (2015) "Inattentional deafness: Visual load leads to time-specific suppression of auditory evoked responses," *J Neurosci*, 35(49):16046–54

14. Lavie N (2005) "Distracted and confused? Selective attention under load," *Trends Cogn Sci*, 9(2):75–82

15. Stothart C, Mitchum A, Yehnert C (2015) "The attentional cost of receiving a cell phone notification," *J Exp Psychol Hum Percept Perform*, 41(4):893–7

16. Gupta R, Hur YJ, Lavie N (2016) "Distracted by pleasure: Effects of positive versus negative valence on emotional capture under load," *Emotion*, 16(3):328–37

17. Lavie N, Ro T, Russell C (2003) "The role of perceptual load in processing distractor faces," *Psychol Sci*, 14(5):510–5

18. Pujol S, Levain JP, Houot H, Petit R, Berthillier M, Defrance J, Lardies J, Masselot C, Mauny F (2014) "Association between ambient noise exposure and school performance of children living in an urban area: A cross-sectional population-based study," *J Urban Health*, 91(2):256–71

19. Halina N, Marsha JE, Hellmana A, Hellströma I, Sörqvista P (2014) "A shield against distraction," *Journal of Applied Research in Memory and Cognition*, 3(1):31–36

20. Pentland A (2012) "The new science of building great teams," *Harvard Business Review*, April Issue

21. Moisala M, Salmela V, Hietajärvi L, Salo E, Carlson S, Salonen O, Lonka K, Hakkarainen K, Salmela-Aro K, Alho K (2016) "Media multitasking is associated with distractibility and increased prefrontal activity in adolescents and young adults," *Neuroimage*, 134:113–21

22. Zabelina DL, O'Leary D, Pornpattananangkul N, Nusslock R, Beeman M (2015) "Creativity and sensory gating indexed by the P50: Selective versus leaky sensory gating in divergent thinkers and creative achievers," *Neuropsychologia*, 69:77–84

23. Mehta R, Zhu RJ, Cheema A (2012) "Is noise always bad? Exploring the effects of ambient noise on creative cognition," *Journal of Consumer Research*, doi: 10.1086/665048

Chapter 8: Mathematics

1. Jänich K (2008) *Topologie* (Springer Lehrbuch), Springer, Heidelberg

2. Arens T, Hettlich F, Karpfinger C, Kockelkorn U, Lichtenegger K, Stachel H (2015) *Mathematik*, Springer, Heidelberg

3. Meyberg K (2003) *Höhere Mathematik 1: Differential- und Integralrechnung, Vektor- und Matrizenrechnung*, Springer, Heidelberg

4. Zeki S, Romaya JP, Benincasa DM, Atiyah MF (2014) "The experience of mathematical beauty and its neural correlates," *Front Hum Neurosci*, doi: 10.3389/fnhum.2014.00068

5.   Siegler RS, Opfer JE (2003) "The development of numerical estimation: Evidence for multiple representations of numerical quantity," *Psychol Sci*, 14(3):237–43

6.   Anobile G, Cicchini GM, Burr DC (2016) "Number as a primary perceptual attribute: A review," *Perception*, 45(1-2):5–31

7.   Arrighi R, Togoli I, Burr DC (2014) "A generalized sense of number," *Proc Biol Sci*, 281(1797)

8.   Nieder A (2016) "The neuronal code for number," *Nat Rev Neurosci*, 17(6):366–82

9.   Pica P, Lemer C, Izard V, Dehaene S (2004) "Exact and approximate arithmetic in an Amazonian indigene group," *Science*, 306(5695): 499–503

10.   Amalric M, Dehaene S (2016) "Origins of the brain networks for advanced mathematics in expert mathematicians," *Proc Natl Acad Sci USA*, 113(18):4909–17

11.   Maruyama M, Pallier C, Jobert A, Sigman M, Dehaene S (2012) "The cortical representation of simple mathematical expressions," *Neuroimage*, 61(4):1444–60

12.   Charness N, Reingold EM, Pomplun M, Stampe DM (2001) "The perceptual aspect of skilled performance in chess: Evidence from eye movements," *Mem Cognit*, 29(8):1146–52

13.   Smalla DA, Loewenstein G, Slovic P (2007) "Sympathy and callousness: The impact of deliberative thought on donations to identifiable and statistical victims," *Organizational Behavior and Human Decision Processes*, 102(2):143–53

Chapter 9: Decisions

1.   www.faz.net/aktuell/feuilleton/apple-ohne-ron-wayne-seine-angst-brachte-ihn-um-dreissig-milliarden-dollar-11558868.html

2.   www.zeit.de/2011/44/P-Wayne

3.   Samanez-Larkin GR, Knutson B (2015) "Decision making in the ageing brain: Changes in affective and motivational circuits," *Nat Rev Neurosci*, 16(5):278–89

4.   De Martino B, Kumaran D, Seymour B, Dolan RJ (2006) "Frames, biases, and rational decision-making in the human brain," *Science*, 313(5787):684–7

5.   Platt ML, Huettel SA (2008) "Risky business: The neuroeconomics of deci-
     sion making under uncertainty," *Nat Neurosci*, 11(4):398–403

6.   Suzuki S, Jensen EL, Bossaerts P, O'Doherty JP (2016) "Behavioral con-
     tagion during learning about another agent's risk-preferences acts on
     the neural representation of decision-risk," *Proc Natl Acad Sci USA*,
     113(14):3755–60

7.   Smith A, Lohrenz T, King J, Montague PR, Camerer CF (2014) "Irrational
     exuberance and neural crash warning signals during endogenous experimen-
     tal market bubbles," *Proc Natl Acad Sci USA*, 111(29):10503–8

8.   Samanez-Larkin GR, Kuhnen CM, Yoo DJ, Knutson B (2010) "Variability in
     nucleus accumbens activity mediates age-related suboptimal financial risk
     taking," *J Neurosci*, 30(4):1426–34

9.   Hsee CK, Ruan B (2016) "The Pandora effect: The power and peril of curio-
     sity," *Psychol Sci*, 27(5):659–66

10. de Berker AO, Rutledge RB, Mathys C, Marshall L, Cross GF, Dolan RJ,
     Bestmann S (2016) "Computations of uncertainty mediate acute stress
     responses in humans," *Nat Commun*, 7:10996

11. Wittmann BC, Bunzeck N, Dolan RJ, Duzel E (2007) "Anticipation of
     novelty recruits reward system and hippocampus while promoting rec-
     ollection," *Neuroimage*, 38(1):194–202

12. Holmes AJ, Hollinshead MO, Roffman JL, Smoller JW, Buckner RL (2016)
     "Individual differences in cognitive control circuit anatomy link sensation
     seeking, impulsivity, and substance use," *J Neurosci*, 36(14):4038–49

Chapter 10: Selection

1.   www.hrk.de/uploads/media/HRK_Statistik_WiSe_2015_16_ webseite_01.
     pdf

2.   Heekeren HR, Marrett S, Ungerleider LG (2008) "The neural systems
     that mediate human perceptual decision making," *Nat Rev Neurosci*,
     9(6):467–79

3.   Aretz W (2015) "Match me if you can: Eine explorative Studie zur Beschrei-
     bung der Nutzung von Tinder." *Journal of Business and Media Psychology*,
     6(1):41–51

4.   Iyengar SS, Lepper MR (2000) "When choice is demotivating: Can one
     desire too much of a good thing?" *J Pers Soc Psychol*, 79(6): 995–1006

5. www.ted.com/talks/sheena_iyengar_choosing_what_to_choose/transcript?language=de

6. Scheibehenne B, Greifeneder R, Todd P (2010) "Can there ever be too many options? A meta-analytic review of choice overload," *Journal of Consumer Research*, 37:409–424

7. Iyengar SS, Lepper MR (2000) "When choice is demotivating: Can one desire too much of a good thing?" *J Pers Soc Psychol*, 79(6):995–1006

8. Chernev A (2003) "Product assortment and individual decision processes," *J Pers Soc Psychol*, 85(1):151–62

9. Huberman G, Iyengar S, Jiang W (2007) "Defined contribution pension plans: Determinants of participation and contributions rates," *Journal of Financial Services Research*, 31(1):1–32

10. Chernev A (2003) "When more is less and less is more: The role of ideal point availability and assortment in consumer choice," *Journal of Consumer Research*, 30(2):170–183

11. Chernev A (2006) "Decision focus and consumer choice among assortments," *Journal of Consumer Research*, 33(6):50–59

12. Scheibehenne B, Greifeneder R, Todd PM (2009) "What moderates the too-much-choice effect?" *Psychology & Marketing*, 26:229–53

13. Inbar Y, Botti S, Hanko K (2011) "Decision speed and choice regret: When haste feels like waste," *Journal of Experimental Social Psychology*, 47(5):533–40

14. Oppewal H, Koelemeijer K (2005) "More choice is better: Effects of assortment size and composition on assortment evaluation," *International Journal of Research in Marketing*, 22(3):45–60

15. Iyengar SS, Wells RE, Schwartz B (2006) "Doing better but feeling worse: Looking for the 'best' job undermines satisfaction," *Psychol Sci*, 17(2):143–50

16. "Entscheiden. Eine Ausstellung uber das Leben im Supermarkt der Moglichkeiten," *Magazin der Arts & Sciences Exhibitions and Publishing GmbH*, Heidelberg 2014

17. Dijksterhuis A, Bos MW, Nordgren LF, van Baaren RB (2006) "On making the right choice: The deliberation-without-attention effect," *Science*, 311(5763):1005–7

18. Lenton AP, Francesconi M (2011) "Too much of a good thing? Variety is confusing in mate choice," *Biol Lett*, 7(4):528–31

19.   Mogilner C, Rudnick T, Iyengar SS (2008) "The mere categorization effect: How the presence of categories increases choosers' perceptions of assortment variety and outcome satisfaction," *Journal of Consumer Research*, 35(8):202–15

20.   "Entscheiden. Eine Ausstellung uber das Leben im Supermarkt der Moglichkeiten," Interview mit Gerd Gigerenzer, p. 60ff., *Magazin der Arts & Sciences Exhibitions and Publishing GmbH*, Heidelberg 2014

Chapter 11: Pigeonholing

1.   Filkuková P, Klempe SH (2014) "Rhyme as reason in commercial and social advertising," *Scand J Psychol*, 54(5):423–31

2.   Tversky A, Kahneman D (1983) "Extensional versus intuitive reasoning: The conjunction fallacy in probability judgment," *Psychological Review*, 90:293–315

3.   Jung K, Shavitt S, Viswanathan M, Hilbe JM (2014) "Female hurricanes are deadlier than male hurricanes," *Proc Natl Acad Sci USA*, 111(24):8782–7

4.   Kutas M, Federmeier KD (2011) "Thirty years and counting: Finding meaning in the N400 component of the event-related brain potential (ERP)," *Annu Rev Psychol*, 62:621–47

5.   Song H, Schwarz N (2008) "If it's hard to read, it's hard to do: Processing fluency affects effort prediction and motivation," *Psychol Sci*, 19(10):986–8

6.   Williams LE, Bargh JA (2008) "Experiencing physical warmth promotes interpersonal warmth," *Science*, 322(5901):606–7

7.   Hicks JA, Cicero DC, Trent J, Burton CM, King LA (2010) "Positive affect, intuition, and feelings of meaning," *J Pers Soc Psychol*, 98(6):967–79

8.   Danziger S, Levav J, Avnaim-Pesso L (2011) "Extraneous factors in judicial decisions," *Proc Natl Acad Sci USA*, 108(17):6889–92

9.   Whitson JA, Galinsky AD (2008) "Lacking control increases illusory pattern perception," *Science*, 322(5898):115–7

10.   Simonov PV, Frolov MV, Evtushenko VF, Sviridov EP (1977) "Effect of emotional stress on recognition of visual patterns," *Aviat Space Environ Med*, 48(9):856–8

11.   Sales SM (1973) "Threat as a factor in authoritarianism: An analysis of archival data," *J Pers Soc Psychol*, 28(1):44–57

12. Gilovich T (1993) *How We Know What Isn't So: The Fallibility of Human Reason in Everyday Life*, The Free Press, New York, p. 16

13. Darley JM, Gross PH (1983) "A hypothesis-confirming bias in labeling effects," *Journal of Personality and Social Psychology*, 44(1):20–33

14. Del Vicario M, Bessi A, Zollo F, Petroni F, Scala A, Caldarelli G, Stanley HE, Quattrociocchi W (2016) "The spreading of misinformation online," *Proc Natl Acad Sci USA*, 113(3):554–9

15. Zollo F, Novak PK, Del Vicario M, Bessi A, Mozetic I, Scala A, Caldarelli G, Quattrociocchi W (2015) "Emotional dynamics in the age of misinformation," *PLoS One*, 10(9):e0138740

16. Mourey JA, Lam BCP, Oyserman D (2015) "Consequences of cultural fluency," *Social Cognition*, 33(4):308–44

17. Norton MC, Smith KR, Østbye T, Tschanz JT, Corcoran C, Schwartz S, Piercy KW, Rabins PV, Steffens DC, Skoog I, Breitner JC, Welsh-Bohmer KA; Cache County Investigators (2010) "Greater risk of dementia when spouse has dementia? The Cache County study," *J Am Geriatr Soc*, 58(5):895–900

18. Khanolkar AR, Ljung R, Talback M, Brooke H, Carlsson S, Mathiesen T, Feychting M (2016) "Socioeconomic position and the risk of brain tumour: A Swedish national population-based cohort study," *J Epidemiol Community Health*, doi: 10.1136/jech-2015-207002

19. Pan W, Altshuler Y, Pentland A (2012) "Decoding social influence and the wisdom of the crowd in financial trading network," *2012 International Conference on Privacy, Security, Risk and Trust and 2012 International Conference on Social Computing*, 203–9, Institute of Electrical and Electronics Engineers (IEEE)

Chapter 12: Motivation

1. Colombo M (2014) "Deep and beautiful. The reward prediction error hypothesis of dopamine," *Stud Hist Philos Biol Biomed Sci*, 45:57–67

2. Pronin E, Olivola CY, Kennedy KA (2008) "Doing unto future selves as you would do unto others: Psychological distance and decision making," *Pers Soc Psychol Bull*, 34(2):224–36

3. Hershfield HE (2011) "Future self-continuity: How conceptions of the future self transform intertemporal choice," *Ann N Y Acad Sci*, 1235:30–43

4. Wilson M, Daly M (2004) "Do pretty women inspire men to discount the future?" *Proc Biol Sci*, 271 Suppl 4:S177–9

5. Eppinger B, Nystrom LE, Cohen JD (2012) "Reduced sensitivity to immediate reward during decision-making in older than younger adults," *PLoS One*, 7(5):e36953

6. Mischel W, Ebbesen EB, Raskoff Zeiss A. (1972) "Cognitive and attentional mechanisms in delay of gratification," *J Pers Soc Psychol*, 21:204–218

7. Mischel W, Ayduk O, Berman MG, Casey BJ, Gotlib IH, Jonides J, Kross E, Teslovich T, Wilson NL, Zayas V, Shoda Y (2010) "'Willpower' over the life span: Decomposing self-regulation," *Soc Cogn Affect Neurosci*, 6(2): 252–56

8. Sturge-Apple ML, Suor JH, Davies PT, Cicchetti D, Skibo MA, Rogosch FA (2016) "Vagal tone and children's delay of gratification: Differential sensitivity in resource-poor and resource-rich environments," *Psychol Sci*, 27(6):885–93

9. Rosenbaum DA, Gong L, Potts CA (2014) "Pre-crastination: Hastening subgoal completion at the expense of extra physical effort," *Psychol Sci*, 25(7):1487–96

10. Bloom M (1999) "The performance effects of pay dispersion on individuals and organizations," *Acad Manage J*, 42(1):25–40

11. Erat S, Gneezy U (2016) "Incentives for creativity," *Exp Econ*, 19(2):269–80

12. Bond RM, Fariss CJ, Jones JJ, Kramer AD, Marlow C, Settle JE, Fowler JH (2012) "A 61-million-person experiment in social influence and political mobilization," *Nature*, 489(7415):295–8

13. Mani A, Loock CM, Rahwan I, Pentland A (2013) "Fostering peer interaction to save energy," *2013 Behavior, Energy, and Climate Change Conference*, Sacramento

14. Murayama K, Matsumoto M, Izuma K, Matsumoto K (2010) "Neural basis of the undermining effect of monetary reward on intrinsic motivation," *Proc Natl Acad Sci USA*, 107(49):20911–6

15. Ariely D, Gneezy U, Loewenstein G, Mazar N (2009) "Large stakes and big mistakes," *Review of Economic Studies*, 76(2):451–69

16. Kuhbandner C, Aslan A, Emmerdinger K, Murayama K (2016) "Providing extrinsic reward for test performance undermines long-term memory acquisition," *Front Psychol*, doi: 10.3389/fpsyg.2016.00079

17. Tricomi E, Fiez JA (2008) "Feedback signals in the caudate reflect goal achievement on a declarative memory task," *Neuroimage*, 41(3):1154–67

Chapter 13: Creativity

1. Kim KH (2011) "The creativity crisis: The decrease in creative thinking scores on the Torrance Tests of Creative Thinking," *Creativity Research Journal*, 23(4):285–95

2. Beaty RE, Benedek M, Silvia PJ, Schacter DL (2016) "Creative cognition and brain network dynamics," *Trends Cogn Sci*, 20(2):87–95

3. Beaty RE, Benedek M, Kaufman SB, Silvia PJ (2015) "Default and executive network coupling supports creative idea production," *Sci Rep*, 5:10964

4. Beaty RE, Benedek M, Wilkins RW, Jauk E, Fink A, Silvia PJ, Hodges DA, Koschutnig K, Neubauer AC (2014) "Creativity and the default network: A functional connectivity analysis of the creative brain at rest," *Neuropsychologia*, 64:92–8

5. Salvi C, Bowden EM (2016) "Looking for creativity: Where do we look when we look for new ideas?" *Front Psychol*, 7:161

6. Mayseless N, Eran A, Shamay-Tsoory SG (2015) "Generating original ideas: The neural underpinning of originality," *Neuroimage*, 116:232–9

7. Hermans EJ, Henckens MJ, Joëls M, Fernández G (2014) "Dynamic adaptation of large-scale brain networks in response to acute stressors," *Trends Neurosci*, 37(6):304–14

8. Rothermel RC (1993) "Mann Gulch fire: A race that couldn't be won," *Gen Tech Rep INT-299*. Ogden, UT: U.S. Department of Agriculture, Forest Service, Intermountain Research Station

9. Zedelius CM, Schooler JW (2015) "Mind wandering 'Ahas' versus mindful reasoning: Alternative routes to creative solutions," *Front Psychol*, 6:834

10. Gasper K, Clore GL (2002) "Attending to the big picture: Mood and global versus local processing of visual information," *Psychol Sci*, 13(1):34–40

11. Bolte A, Goschke T, Kuhl J (2003) "Emotion and intuition: Effects of positive and negative mood on implicit judgments of semantic coherence," *Psychological Science*, 14(5):416–21

12. Kounios J, Beeman M (2014) "The cognitive neuroscience of insight," *Annu Rev Psychol*, 65:71–93

13. t3n.de/magazin/yo-app-239034/

14. Slepian ML, Weisbuch M, Rutchick AM, Newman LS, Ambady N (2010) "Shedding light on insight: Priming bright ideas," *J Exp Soc Psychol*, 46(4):696–700

15. Oppezzo M, Schwartz DL (2014) "Give your ideas some legs: The positive effect of walking on creative thinking," *J Exp Psychol Learn Mem Cogn*, 40(4):1142–52

16. Trope Y, Liberman N (2010) "Construal-level theory of psychological distance," *Psychol Rev*, 117(2):440–63

17. Olguin OD, Waber BN, Kim T, Mohan A, Ara K, Pentland A (2009) "Sensible organizations: Technology and methodology for automatically measuring organizational behavior," *IEEE Trans Syst Man Cybern B Cybern*, 39(1):43–55

18. Pentland A (2012) "The new science of building great teams," *Harvard Business Review*, April Issue

19. Saggar M, Quintin EM, Kienitz E, Bott NT, Sun Z, Hong WC, Chien YH, Liu N, Dougherty RF, Royalty A, Hawthorne G, Reiss AL (2015) "Pictionary-based fMRI paradigm to study the neural correlates of spontaneous improvisation and figural creativity," *Sci Rep*, 5:10894

Chapter 14: Perfectionism

1. straightsets.blogs.nytimes.com/2010/06/23/logistics-are-put-to-the-test-at-wimbledon/

2. Brickenkamp R, Schmidt-Atzert L, Liepmann D (2014) "Test d2–Revision–Aufmerksamkeits- und Konzentrationstest (d2-R)," *Dorsch–Lexikon der Psychologie*, 17th ed., p. 1648

3. Hoffmann S, Beste C (2015) "A perspective on neural and cognitive mechanisms of error commission," *Front Behav Neurosci*, doi: 10.3389/fnbeh.2015.00050

4. van Veen V, Carter CS (2006) "Error detection, correction, and prevention in the brain: A brief review of data and theories," *Clin EEG Neurosci*, 37(4):330–5

5. Debener S, Ullsperger M, Siegel M, Fiehler K, von Cramon DY, Engel AK (2005) "Trial-by-trial coupling of concurrent electroencephalogram and functional magnetic resonance imaging identifies the dynamics of performance monitoring," *J Neurosci*, 25(50):1730–7

6. Rodriguez-Fornells A, Kurzbuch AR, Münte TF (2002) "Time course of error detection and correction in humans: Neurophysiological evidence," *J Neurosci*, 22(22):9990–6

7.   Perri RL, Berchicci M, Lucci G, Spinelli D, Di Russo F (2016) "How the brain prevents a second error in a perceptual decision-making task," *Sci Rep*, doi: 10.1038/srep32058

8.   www.wired.com/2012/09/deep-blue-computer-bug/

9.   Mehl K (2015) "Warum wir Fehler machen und benötigen," in: *Fehler: Ihre Funktion im Kontext individueller und gesellschaftlicher Entwicklung*, p. 129–40, Waxmann Verlag, Münster

10.   Kapur M (2014) "Productive failure in learning math," *Cogn Sci*, 38(5):1008–22

11.   Butler AC, Karpicke JD, Roediger HL 3rd (2007) "The effect of type and timing of feedback on learning from multiple-choice tests," *J Exp Psychol Appl*, 13(4):273–81

12.   Affrunti NW, Geronimi EM, Woodruff-Borden J (2015) "Language of per-fectionistic parents predicting child anxiety diagnostic status," *J Anxiety Disord*, 30:94–102